Max Plus at Work

PRINCETON SERIES IN APPLIED MATHEMATICS

Max Plus at Work

*Modeling and Analysis of Synchronized
Systems: A Course on Max-Plus Algebra and
Its Applications*

**Bernd Heidergott
Geert Jan Olsder
Jacob van der Woude**

PRINCETON UNIVERSITY PRESS

PRINCETON AND OXFORD

Published by Princeton University Press, 41 William Street,
Princeton, New Jersey 08540
In the United Kingdom: Princeton University Press,
3 Market Place, Woodstock, Oxfordshire OX20 1SY

Library of Congress Cataloging-in-Publication Data

Heidergott, Bernd.
Max Plus at work : modeling and analysis of synchronized systems : a course on
Max-Plus algebra and its applications / Bernd Heidergott, Geert Jan Olsder,
Jacob van der Woude.
     p. cm.—(Princeton series in applied mathematics)
     Includes bibliographical references and index.
     ISBN-13: 978-0-691-11763-8 (acid-free paper)
     ISBN-10: 0-691-11763-2 (acid-free paper)
     1. Matrices—Textbooks. 2. System theory—Textbooks. I. Olsder, Geert Jan. II.
Woude, J. W. van der. III. Title.
     QA188.H445 2006
     512—dc22   2005048690

British Library Cataloging-in-Publication Data is available

The publisher would like to acknowledge the authors of this volume for providing
the camera-ready copy from which this book was printed.

Printed on acid-free paper.

pup.princeton.edu

Printed in the United States of America

10 9 8 7 6 5 4 3 2 1

Contents

Preface

This textbook is concerned with sequences of events, where events are viewed as sudden changes in a process to be studied. Some examples of events are a message arrives, a train leaves the station, and a door opens. This book deals with the modeling, analysis, and timing of such events, all subject to synchronization constraints. These constraints are relations that exist between the events, such as a message must have been sent before it can arrive and a certain train should not depart before another train has arrived (in order to allow the changeover of passengers).

Apart from the introductory Chapter 0, the book consists of three parts. Part I (Chapters 1–6) deals with max-plus algebra, i.e., an algebra in which the evolution of events can be described in a convenient way. Part II (Chapters 7–10) covers two specific applications, both related to timetable design for railway networks. Part III (Chapters 11–13) deals with various extensions. Later on in this Preface, we give a brief description of each chapter.

The level of the book is last-year undergraduate student mathematics. The book will be of interest for applied mathematicians, operations researchers, econometricians, and civil, electrical, and mechanical engineers with quantitative backgrounds. Most important is a basic knowledge of conventional algebra and matrix calculus, with the addition of some knowledge of system theory (or recurrence relations). Some knowledge of stochastic processes is required for Chapter 11 only. No prior knowledge of graph theory and the modeling tool of Petri nets is required.

Each chapter can be taught conveniently in two hours. The only exceptions to this rule are Chapter 3 (which requires at least three hours) and Chapter 10 (for which one hour is enough). With two hours per week, this would mean that fourteen weeks would suffice to teach the whole book. Of course, depending on interest, some chapters can be studied in more depth, requiring more than two hours, while others can be skipped or can be treated superficially in one hour. A minimum course would cover the material of Chapters 0, 1, 2, 7 and 8. From a mathematical point of view, Chapter 3 is probably the most challenging in presenting key results. All chapters contain an exercise section as well as a notes section with suggestions for further reading and sometimes other remarks.

On the timescale of mathematical evolution, max-plus algebra and its applications are a recent phenomenon. Originally, some results were published in journals (e.g., [30], [43]), and even earlier traces exist. The first major leap forward in this algebra was the appearance of the book [31] in 1979. The next book, written from a system-theoretical perspective, was [5]. It can be viewed as a product of the "French school." Neither book can be called a textbook. The current book is believed to be the first textbook in the area of max-plus algebra and its applications. Other textbooks with introductions to the much wider area of discrete event systems do exist (e.g., [20]). Though the book at hand concentrates on applications related to timetables, other realistic applications do exist, for instance, in the areas of production lines and network calculus. For references the reader is referred to the notes sections of the chapters of this book.

A brief description of the book now follows. Chapter 0 gives an overview of some concepts to be dealt with and some problems to be solved. These concepts, problems, and solutions are elucidated by means of a simple academic example.

Part I, consisting of Chapters 1–6, contains the core of the theory and forms the basis for Parts II and III. Chapter 1 introduces max-plus algebra, which can be viewed as a mutation of conventional algebra. In max-plus algebra, the operations max (being maximization) and plus (being addition) play a fundamental role. Vectors, matrices, and the notion of linearity are introduced within this new algebra. The "heaps of pieces" point of view provides a first application. Chapter 2 deals with eigenvalues and eigenvectors of matrices in max-plus algebra and their graph-theoretic interpretations. Sets of linear equations are studied also. In Chapters 3 and 4, we explore linear systems in max-plus algebra and study their behavior in terms of throughput, growth rate, and periodicity. Various concepts are introduced, such as (ir)reducibility of the system matrix or its graph having a sunflower shape. Chapters 5 and 6 deal with numerical procedures to calculate characteristic quantities of a matrix, such as the eigenvalue and its extension, the so-called generalized eigenmode. The three procedures treated are named Karp's algorithm, the power algorithm, and Howard's algorithm. The last one especially is very efficient for large-scale matrices.

In Part II we examine Petri nets and real-life applications, mainly drawn from everyday railway issues in the Netherlands. The subject of Chapter 8 is a study of the timetable for the whole of the Dutch railway system. Since the detailed description of the whole network would obscure the methods used, we decided to describe a subnetwork of dimension 24. This subnetwork contains all the details of how to arrive at a max-plus model starting from line and synchronization data as provided by the railway company. Petri nets form a very convenient intermediate tool to connect this data to max-plus models. Therefore, the preceding chapter, Chapter 7, is devoted to the introduction of Petri nets. Chapter 9 deals with delay propagation and various stability measures for railway networks. We also discuss issues such as an optimal allocation of trains and their ordering. The application of Chapter 10 concerns a series of railway tunnels for which capacity issues are discussed. Having come to the end of Part II, the reader should be able to conclude that max plus is at work indeed!

In Part III we explore some extensions of the theory treated so far, and this section can be read independently of Part II. Chapter 11 deals with various stochastic extensions. The subject of Chapter 12 is min-max-plus systems, which are max-plus systems, described by the max and plus operation, to which the min operation, being minimization, is added. Thus, a larger class of problems can be modeled. The relationship to the theory of nonnegative matrices and nonexpansive mappings is indicated. Lastly, Chapter 13 deals with continuous flows on networks, which, theoretically speaking, can be viewed as the continuous counterpart of discrete events on networks. Though we had been thinking about including a chapter on the control of input/output systems, we decided not to do so. The subject concerned requires a background in residuation theory that is beyond the scope of this book. Those interested are referred to [29] and the references therein.

The book ends with a bibliography, a list of frequently used symbols, and an index.

For the preparation of this book we should like to acknowledge the help and contributions of various colleagues. The second author would like to thank CNRS (Centre National de la Recherche Scientifique) in the person of Pierre Bernhard, for allowing him to work on this book project while spending a sabbatical at I3S, Sophia Antipolis, France. Carl Schneider deserves thanks for his help drawing some of the figures. Rob Goverde gave us insight into the intricacies of the software package PETER. Anton Stoorvogel, Katarína Cechlárová, Jean-Louis Boimond, Niek Tholen, and Ton van den Boom were so kind as to read through a preliminary version of this book; they came up with many valuable comments. Besides, both Carl Schneider and Anton Stoorvogel were helpful in solving several LATEXpuzzles. We also thank our former PhD students Remco de Vries, Hans Braker, Erik van Bracht, Subiono, Robert-Jan van Egmond, Antoine de Kort, and Gerardo Soto y Koelemeijer for the many discussions we had. Furthermore, Stéphane Gaubert provided extra references, and both he and Pierre Bernhard gave additional comments. Also thanks to the two, originally anonymous reviewers Jean-Pierre Quadrat and Bart De Schutter, as invited by the publisher, for their constructive remarks and criticism. Finally, we thank our universities for providing the right atmosphere.

Dear reader: We hope that you will enjoy reading this book as much as we enjoyed writing it. Have a good max-plus trip. Bon voyage!

The authors   Delft, The Netherlands, January 2005

Bernd Heidergott, Vrije Universiteit, Amsterdam.
Geert Jan Olsder, Delft University of Technology, Delft.
Jacob van der Woude, Delft University of Technology, Delft.

Chapter Zero

Prolegomenon

In this book we will model, analyze, and optimize phenomena in which the order of events is crucial. The timing of such events, subject to synchronization constraints, forms the core. This zeroth chapter can be viewed as an appetizer for the other chapters to come.

0.1 INTRODUCTORY EXAMPLE

Consider a simple railway network between two cities, each with a station, as indicated in Figure 0.1. These stations are called S_1 and S_2, respectively, and are connected by two tracks. One track runs from S_1 to S_2, and the travel time for a train along this track is assumed to be 3 time units. The other track runs from S_2 to S_1, and a voyage along this track lasts 5 time units. Together, these two tracks form a circuit. Trains coming from S_1 and arriving in S_2 will return to S_1 along the other track, and trains that start at S_2 will, after having visited S_1, come back to S_2. Apart from these two tracks, two other tracks, actually circuits, exist, connecting the suburbs of a city with its main station. A round trip along these tracks lasts 2 units of time for the first city and 3 time units for the second city. Of course, local stations exist in these suburbs, but since they will not play any role in the problem, they are not indicated. We want to design a timetable subject to the following criteria:

- The travel times of the trains along each of the tracks are fixed (and given).

- The frequency of the trains (i.e., the number of departures per unit of time) must be as high as possible.

- The frequency of the trains must be the same along all four tracks, yielding a timetable with regular departure times.

- The trains arriving at a station should wait for each other in order to allow the changeover of passengers.

- The trains at a station depart the station as soon as they are allowed.

We will start with a total number of four trains in the model, one train on each of the outer circuits and two trains on the inner circuit. The departure time of the two trains at station S_1, one in the direction of S_2 and the other one to serve the suburbs, will be indicated by x_1. These two trains depart at the same time because of the requirement of the changeover of passengers and that trains depart as soon

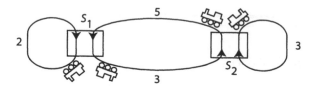

Figure 0.1: The railway network. The numbers along the tracks refer to travel times.

as possible. Similarly, x_2 is the common departure time of the two trains at S_2. Together, the departure times are written as the vector $x \in \mathbb{R}^2$. The first departure times during a day, in the early morning, will be given by $x(0)$. The trains thereafter leave at the time instants given by the two elements of the vector $x(1)$ and so on. The kth departure times are indicated by $x(k-1)$. These departures are called *events* in the model. Because of the rules given, it follows that

$$\begin{aligned} x_1(k+1) &\geq x_1(k) + a_{11} + \delta, \\ x_1(k+1) &\geq x_2(k) + a_{12} + \delta. \end{aligned} \tag{0.1}$$

The quantities a_{ij} denote the travel time from the station indicated by the second subscript (S_j) to the station indicated by the first subscript (S_i), and δ denotes the time reserved for the passengers to change from one train to the other. Without loss of generality, δ can be thought of being part of the travel time. In other words, the travel time can be defined as the actual travel time to which the changeover time, or transfer time, has been added. Hence, from now on it will be assumed that $\delta = 0$. Substituting $a_{11} = 2$ and $a_{12} = 5$, it follows that

$$x_1(k+1) \geq \max(x_1(k) + 2, x_2(k) + 5).$$

Similarly, the departure times at S_2 must satisfy

$$x_2(k+1) \geq \max(x_1(k) + 3, x_2(k) + 3).$$

Since the frequency of the departures must be as high as possible and the trains depart as soon as possible, the inequalities in the latter two expressions will, in fact, have to be equalities, which leads to

$$\begin{aligned} x_1(k+1) &= \max(x_1(k) + 2, x_2(k) + 5), \\ x_2(k+1) &= \max(x_1(k) + 3, x_2(k) + 3). \end{aligned} \tag{0.2}$$

Then, if the initial departure times $x(0)$ are given, all future departure times are uniquely determined. If for instance $x_1(0) = x_2(0) = 0$, then the sequence $x(k)$, for $k = 0, 1, \ldots$, becomes

$$\begin{pmatrix} 0 \\ 0 \end{pmatrix}, \begin{pmatrix} 5 \\ 3 \end{pmatrix}, \begin{pmatrix} 8 \\ 8 \end{pmatrix}, \begin{pmatrix} 13 \\ 11 \end{pmatrix}, \begin{pmatrix} 16 \\ 16 \end{pmatrix}, \ldots \tag{0.3}$$

Compare this sequence with the following one, obtained if the initial departure times are $x_1(0) = 1$ and $x_2(0) = 0$ (i.e., the first trains at S_2, one in each direction, still leave at time 0, but the first trains at S_1 now leave at time 1),

$$\begin{pmatrix} 1 \\ 0 \end{pmatrix}, \begin{pmatrix} 5 \\ 4 \end{pmatrix}, \begin{pmatrix} 9 \\ 8 \end{pmatrix}, \begin{pmatrix} 13 \\ 12 \end{pmatrix}, \begin{pmatrix} 17 \\ 16 \end{pmatrix}, \ldots \tag{0.4}$$

With the interdeparture time being the time duration between two subsequent departures along the same track, both sequences have the same *average* interdeparture time equal to 4, but the second sequence has exactly this interdeparture time, whereas the first sequence has it only on average (the average of the interdeparture times 3 and 5). If these sequences were real timetable departures, then most people would prefer the second timetable since it is regular.

A question that might arise is whether it would be possible to have a "faster" timetable (i.e., a timetable with a smaller average interdeparture time) by choosing appropriate initial departure times? The answer is no. The reason is that the time duration for a train to go around on the inner circuit is equal to 8, and there are two trains on this circuit. Hence, the average interdeparture time can never be smaller than $8/2 = 4$.

As a direct generalization of (0.2), one can study

$$x_i(k + 1) = \max(x_1(k) + a_{i1}, x_2(k) + a_{i2}, \ldots, x_n(k) + a_{in}), \qquad (0.5)$$

for $i = 1, 2, \ldots, n$, and in fact, the study of these equations will be the central theme of this book. The x_i's could, for instance, be the departure times in a more general railway network, with the explicit possibility that some of the terms in the right-hand side are not present. In the terminology of train networks this would mean, as in reality, that there are no direct tracks between some of the stations. This absence of terms is solved by allowing the value $-\infty$ for the corresponding a_{ij}-quantities. The value $-\infty$ will never contribute to the max operation, and thus there is no need to change the notation in (0.5). In the parlance of train networks, if a direct connection between stations S_j and S_i does not exist, then simply define its corresponding travel time to be equal to $-\infty$, i.e., $a_{ij} = -\infty$. This is done for mathematical convenience. At first sight one might be tempted to set $a_{ij} = +\infty$ for a nonexisting track. Setting $a_{ij} = +\infty$, however, refers to an existing track with an extremely high travel time, and a term with such an a_{ij}-quantity in a max-plus expression will dominate all other terms.

We can draw a directed graph based on (0.5). Such a graph has n nodes, one node corresponding to each x_i, and a number of directed arcs (i.e., arrows), one arc from node j to node i for each $a_{ij} \neq -\infty$. More details follow in Chapter 2.

0.2 ON THE NOTATION

Equation (0.5) can be written more compactly as

$$x_i(k + 1) = \max_{j=1,2,\ldots,n} (a_{ij} + x_j(k)), \qquad i = 1, 2, \ldots, n. \qquad (0.6)$$

Many readers will be familiar with linear recurrence relations of the form

$$z_i(k + 1) = \sum_{j=1}^{n} c_{ij} z_j(k), \qquad i = 1, 2, \ldots, n.$$

For conceptual reasons the above equation will also be written as

$$z_i(k + 1) = \sum_{j=1}^{n} c_{ij} \times z_j(k), \qquad i = 1, 2, \ldots, n, \qquad (0.7)$$

and the reader will now immediately notice the resemblance between (0.6) and
(0.7). The "only" difference between the two expressions is that the maximization
and addition in the first equation are replaced by the addition and multiplication,
respectively, in the second equation. In order to make this resemblance more clear
we will change the notation of the operations in (0.6). For (0.6) we will henceforth
write

$$x_i(k+1) = \bigoplus_{j=1}^{n} a_{ij} \otimes x_j(k), \qquad i = 1, 2, \ldots, n. \qquad (0.8)$$

For the pronunciation of the symbol $\oplus$ and related ones, which will be introduced
later, the reader is referred to Table 0.1.

symbol	pronunciation	symbol	pronunciation
$\bigoplus$	big o-plus	$\oplus$	o-plus
$\bigotimes$	big o-times	$\otimes$	o-times
$\bigoplus'$	big o-plus prime	$\otimes'$	o-times prime

Table 0.1: The pronunciation of some symbols.

If the circles around $\bigoplus$ and $\otimes$ in (0.8) were omitted, we would get equations of
type (0.7), be it that the summations still are different in notation. To be very clear,
the evolutions with respect to the parameter k of the processes characterized by
(0.6) and (0.8) will be the same and will generally be different from the evolution
resulting from (0.7). In conventional linear algebra the scalar equations (0.7) can
be written in vector form as

$$z(k+1) = Cz(k).$$

In the same way, we will write (0.5) in vector form as

$$x(k+1) = A \otimes x(k), \qquad (0.9)$$

with $\otimes$ to indicate that the underlying process is not described in terms of the conventional linear algebra (upon which (0.7) is based). The algebra underlying equations of the form (0.5) and (0.9) is called *max-plus algebra*. The relation between
$\bigoplus$ in (0.8) and $\oplus$ in Table 0.1 is clarified by

$$\bigoplus_{j=1}^{n} (a_{ij} \otimes x_j) = (a_{i1} \otimes x_1) \oplus (a_{i2} \otimes x_2) \oplus \cdots \oplus (a_{in} \otimes x_n)$$

for any $i = 1, 2, \ldots, n$.

The reason for wanting to emphasize the resemblance between the notation in
conventional algebra and that of max-plus algebra is partly historical. More important, many well-known concepts in conventional linear algebra can be carried over
to max-plus algebra, which can be emphasized with a similar notation. The next
section on eigenvectors will make this particularly clear, where the evolution of the
state of model (0.9) will be considered in more detail.

For $k = 1$ we get $x(1) = A \otimes x(0)$, and for $k = 2$

$$\begin{aligned} x(2) &= A \otimes x(1) \\ &= A \otimes (A \otimes x(0)) \\ &= (A \otimes A) \otimes x(0) \\ &= A^{\otimes 2} \otimes x(0). \end{aligned}$$

The associative property, together with others, will be discussed in Chapter 1. Instead of $A \otimes A$, we have simply written $A^{\otimes 2}$, where the $\otimes$ symbol in the exponent indicates a matrix power in max-plus algebra. Continuing along these lines, we get

$$\begin{aligned} x(3) &= A \otimes x(2) \\ &= A \otimes (A^{\otimes 2} \otimes x(0)) \\ &= (A \otimes A^{\otimes 2}) \otimes x(0) \\ &= A^{\otimes 3} \otimes x(0), \end{aligned}$$

and in general,

$$\begin{aligned} x(k) &= A \otimes x(k-1) \\ &= A \otimes (A^{\otimes(k-1)} \otimes x(0)) \\ &= (A \otimes (A^{\otimes(k-1)})) \otimes x(0) \\ &= (\underbrace{A \otimes A \otimes \cdots \otimes A}_{k \text{ times}}) \otimes x(0) \\ &= A^{\otimes k} \otimes x(0). \end{aligned}$$

The matrices $A^{\otimes 2}$, $A^{\otimes 3}$, ... can be calculated directly.

As an example, let us consider the example in Section 0.1 once more. The equations governing the departure times were given by (0.2) or, equivalently, by (0.9) with

$$x(k) = \begin{pmatrix} x_1(k) \\ x_2(k) \end{pmatrix} \quad \text{and} \quad A = \begin{pmatrix} 2 & 5 \\ 3 & 3 \end{pmatrix}. \tag{0.10}$$

Then we get (details on vector and matrix multiplication follow in Chapter 1)

$$A^{\otimes 2} \begin{pmatrix} 2 & 5 \\ 3 & 3 \end{pmatrix} \otimes \begin{pmatrix} 2 & 5 \\ 3 & 3 \end{pmatrix} = \begin{pmatrix} (2 \otimes 2) \oplus (5 \otimes 3) & (2 \otimes 5) \oplus (5 \otimes 3) \\ (3 \otimes 2) \oplus (3 \otimes 3) & (3 \otimes 5) \oplus (3 \otimes 3) \end{pmatrix}$$

$$= \begin{pmatrix} \max(2+2, 5+3) & \max(2+5, 5+3) \\ \max(3+2, 3+3) & \max(3+5, 3+3) \end{pmatrix} = \begin{pmatrix} 8 & 8 \\ 6 & 8 \end{pmatrix}.$$

If the initial departure times are $x_1(0) = 1$ and $x_2(0) = 0$, then we can directly calculate that

$$x(2) = A^{\otimes 2} \otimes x(0) = \begin{pmatrix} 8 & 8 \\ 6 & 8 \end{pmatrix} \otimes \begin{pmatrix} 1 \\ 0 \end{pmatrix}$$

$$= \begin{pmatrix} \max(8+1, 8+0) \\ \max(6+1, 8+0) \end{pmatrix} = \begin{pmatrix} 9 \\ 8 \end{pmatrix}, \tag{0.11}$$

which is in complete agreement with the corresponding result in (0.4).

In general, the entry of the matrix $A^{\otimes 2}$ in row i and column j is given by

$$[A^{\otimes 2}]_{ij} = \bigoplus_{l=1}^{n} a_{il} \otimes a_{lj} = \max_{l=1,\ldots,n} (a_{il} + a_{lj}).$$

The quantity $[A^{\otimes 2}]_{ij}$ can be interpreted as the maximum (with respect to l) over all connections (think of the directed graph representing A) from node j to node i via node l. In terms of the train example, this maximum refers to the maximum of all travel times from j to i via l. More generally, $[A^{\otimes k}]_{ij}$ will denote the maximum travel time from j to i via $k-1$ intermediate nodes. In graph-theoretical terms one speaks about paths of length k, instead of connections via $k-1$ intermediate nodes, starting at node j and ending at node i.

0.3 ON EIGENVALUES AND EIGENVECTORS

We consider the notion of eigenvalue and eigenvector in max-plus algebra. Given a square matrix A of size $n \times n$, assume that

$$A \otimes v = \lambda \otimes v, \tag{0.12}$$

where λ is a scalar and v is an n-vector such that not all of its components are equal to $-\infty$. The notation $\lambda \otimes v$ refers to an n-vector whose ith element equals $\lambda \otimes v_i$, i.e., is equal to $\lambda + v_i$. If λ and v are as described in (0.12), then λ is called an *eigenvalue* of matrix A and v a corresponding *eigenvector*.

If $x(0)$ is an eigenvector of A corresponding to the eigenvalue λ, then the solution of (0.9) can be written as

$$x(1) = A \otimes x(0) = \lambda \otimes x(0),$$

$$x(2) = A \otimes x(1) = A \otimes (\lambda \otimes x(0)) = \lambda^{\otimes 2} \otimes x(0),$$

and in general,

$$x(k) = \lambda^{\otimes k} \otimes x(0), \qquad k = 0, 1, 2, \ldots.$$

Note that the numerical evaluation of $\lambda^{\otimes k}$ in max-plus algebra is equal to $k \times \lambda$ in conventional algebra. It is easily seen that the eigenvector is not unique. If the same constant is added to all elements of $x(0)$, then the resulting vector will again be an eigenvector. This is reminiscent of the situation in conventional algebra in which the eigenvectors are determined up to a multiplicative factor.

As an example, observe that

$$\begin{pmatrix} 2 & 5 \\ 3 & 3 \end{pmatrix} \otimes \begin{pmatrix} 1+h \\ h \end{pmatrix} = 4 \otimes \begin{pmatrix} 1+h \\ h \end{pmatrix}$$

for arbitrary h. Thus, it is seen that the matrix from the example in Section 0.1 has eigenvalue 4. Also compare the sequence resulting for $h = 0$ with (0.4). Moreover, if the initial condition of the corresponding system (0.10) happens to be an eigenvector, then the evolution of the states according to (0.10) leads directly to a regular timetable. Equation (0.11) can be written as

$$x(2) = \begin{pmatrix} 8 & 8 \\ 6 & 8 \end{pmatrix} \otimes \begin{pmatrix} 1 \\ 0 \end{pmatrix} = \begin{pmatrix} 9 \\ 8 \end{pmatrix} = \lambda^{\otimes 2} \otimes \begin{pmatrix} 1 \\ 0 \end{pmatrix}.$$

The existence of eigenvalues and eigenvectors will be studied in more depth in Chapter 2. In Chapters 5 and 6 we will give computational schemes for their calculation.

0.4 SOME MODELING ISSUES

Suppose that the management of the railway company in Section 0.1 decides to buy an extra train in order to possibly speed up the network's behavior (i.e., to obtain a timetable with an average interdeparture time less than $\lambda = 4$). On which circuit should this extra train run? Suppose that the extra train is placed on the track from S_1 to S_2, just outside station S_1 and at the moment that train number k has already left in the direction of S_2. Hence, train number k is in front of the newly added train. If this train number k is renumbered as the $(k-1)$st train and the newly added train gets number k, then the model that yields the smallest possible departure times is given by

$$\begin{aligned} x_1(k+1) &= \max(x_1(k) + 2, x_2(k) + 5), \\ x_2(k+1) &= \max(x_1(k-1) + 3, x_2(k) + 3), \end{aligned} \tag{0.13}$$

which can be rewritten as a first-order recurrence relation by introducing an auxiliary variable x_3 with $x_3(k+1) \stackrel{\text{def}}{=} x_1(k)$ as follows:

$$\begin{pmatrix} x_1(k+1) \\ x_2(k+1) \\ x_3(k+1) \end{pmatrix} = \begin{pmatrix} 2 & 5 & -\infty \\ -\infty & 3 & 3 \\ 0 & -\infty & -\infty \end{pmatrix} \otimes \begin{pmatrix} x_1(k) \\ x_2(k) \\ x_3(k) \end{pmatrix}. \tag{0.14}$$

In order to interpret this in a different way, one can think of the auxiliary variable x_3 as the departure time at an auxiliary station S_3 situated on the track from S_1 to S_2, just outside S_1, such that the travel time between S_1 and S_3 equals 0 and the travel time from S_3 to S_2 is 3.

There are, of course, other places where the auxiliary station could be situated, for example, somewhere on the inner circuit or on one of the two outer circuits. If, instead of having S_3 neighboring S_1 as above, one could situate S_3 just before S_2, still on the track from S_1 to S_2, then the equations become

$$\begin{pmatrix} x_1(k+1) \\ x_2(k+1) \\ x_3(k+1) \end{pmatrix} = \begin{pmatrix} 2 & 5 & -\infty \\ -\infty & 3 & 0 \\ 3 & -\infty & -\infty \end{pmatrix} \otimes \begin{pmatrix} x_1(k) \\ x_2(k) \\ x_3(k) \end{pmatrix}. \tag{0.15}$$

Or, with S_3 just after S_2 on the track towards S_1,

$$\begin{pmatrix} x_1(k+1) \\ x_2(k+1) \\ x_3(k+1) \end{pmatrix} = \begin{pmatrix} 2 & -\infty & 5 \\ 3 & 3 & -\infty \\ -\infty & 0 & -\infty \end{pmatrix} \otimes \begin{pmatrix} x_1(k) \\ x_2(k) \\ x_3(k) \end{pmatrix}. \tag{0.16}$$

Each of the three models (0.14), (0.15), and (0.16) essentially describes the same speedup of the network's behavior. It will come as no surprise that the eigenvalues of the three corresponding system matrices are identical. A little exercise shows that these eigenvalues all equal 3. That the eigenvalues, or average interdeparture times, cannot be smaller than 3 is easy to understand since the outer circuit at S_2 has one train and the travel time equals 3. On the inner circuit the average interdeparture time cannot be smaller than 8/3 (i.e., the total travel time on this circuit divided by the number of trains on it). Apparently, the outer circuit at S_2 has become the bottleneck now. A small calculation will show that eigenvectors corresponding to models (0.14), (0.15), and (0.16) are

$$\begin{pmatrix} 0 \\ -2 \\ -3 \end{pmatrix}, \quad \begin{pmatrix} 0 \\ -2 \\ 0 \end{pmatrix}, \text{ and } \begin{pmatrix} 0 \\ 1 \\ -2 \end{pmatrix},$$

respectively. Since eigenvectors are determined up to the addition of a constant, the eigenvectors given above are scaled in such a way that the first element is equal to zero. At first sight, one may be surprised about the fact that the departure times at S_1 and S_2 differ in the latter two cases. For the first (and second) model, the departure times at S_1 are $0, 3, 6, 9, \ldots$, and for S_2 they are $-2, 1, 4, 7, \ldots$. For the third model these sequences are $0, 3, 6, 9, \ldots$ and $1, 4, 7, 10, \ldots$, respectively. Thus, one notices that the kth departure time at S_2 of model (0.15) coincides with the $(k-1)$st departure time at S_2 of model (0.16). Apparently the geographical shift of station S_3 on the inner circuit, from just before S_2 to just after it, causes a shift in the counting of the departures and their times.

Models (0.14) and (0.15) can be obtained from one another by means of a coordinate transformation in the max-plus algebra sense. With $A_{\text{eqn (0.14)}}$ being the matrix from (0.14) and similarly for $A_{\text{eqn (0.15)}}$, it is a straightforward calculation to show that

$$T^{\otimes -1} \otimes A_{\text{eqn (0.14)}} \otimes T = A_{\text{eqn (0.15)}}, \qquad T^{\otimes -1} \otimes T = E,$$

where E denotes the identity matrix in max-plus algebra (i.e., zeros on the diagonal and $-\infty$'s elsewhere) and

$$T = \begin{pmatrix} 0 & -\infty & -\infty \\ -\infty & 0 & -\infty \\ -\infty & -\infty & -3 \end{pmatrix}, \qquad T^{\otimes -1} = \begin{pmatrix} 0 & -\infty & -\infty \\ -\infty & 0 & -\infty \\ -\infty & -\infty & 3 \end{pmatrix}.$$

A similar coordinate transformation does not exist between models (0.15) and (0.16). This is left as an exercise. The reader should perhaps at this point already be warned that, in contrast to the transformation just mentioned, the inverse of a matrix does in general not exist in max-plus algebra.

0.5 COUNTER AND DATER DESCRIPTIONS

Traditionally, the introduction of a max-plus system is as given in this chapter. Other approaches to describe the same kind of phenomena exist. Define $\kappa_i(\chi)$ as the number of trains in a certain direction that have left station S_i up to and including time χ. Be aware of the fact that here the argument χ refers to time and κ to a counter. In the notation $x_i(k)$ it is the other way around, since there x_i is the time at which an event takes place and k is the counter. The model of the example in Section 0.1 can be written in terms of the κ-variables as

$$\begin{aligned}
\kappa_1(\chi) &= \min(\kappa_1(\chi - 2) + 1, \kappa_2(\chi - 5) + 1), \\
\kappa_2(\chi) &= \min(\kappa_1(\chi - 3) + 1, \kappa_2(\chi - 3) + 1).
\end{aligned} \tag{0.17}$$

At station S_1, for instance, the number of departures up to time χ cannot be more than one plus the number of departures from the station S_1 2 time units ago, due to the outer loop, and also not more than one plus the number of departures from station S_2 5 time units ago, due to the inner loop. Recall that a departure here means a departure in both directions (one on the inner loop, one on the outer loop) simultaneously.

Equations (0.17) are equations in so-called min-plus algebra, due to the fact that the minimization and addition play a role in the evolution. By augmenting the state space (i.e., introducing auxiliary κ-variables), (0.17) can be written as a set of first-order recurrence relations, which symbolically can be written as

$$\kappa(\chi) = B \otimes' \kappa(\chi - 1), \tag{0.18}$$

where the matrix B is a square matrix. The symbol $\otimes'$ (see Table 0.1 for its pronunciation) indicates that we are working in min-plus algebra. While in max-plus algebra the "number" $-\infty$ was introduced to characterize nonexisting connections, in min-plus algebra this role is taken over by $+\infty$. Indeed, $\min(a, +\infty) = a$ for any finite a. Equations in min-plus algebra are referred to as counter equations and equations in max-plus algebra as dater equations.

0.6 EXERCISES

1. Define the max-plus product $A \otimes B$ of two matrices A and B of size $n \times n$, as (see Chapter 1 for the general definition)

$$[A \otimes B]_{ij} = \bigoplus_{l=1}^{n} a_{il} \otimes b_{lj} = \max_{1 \le l \le n} (a_{il} + b_{lj}),$$

and compute $A^{\otimes 3}$ as $A \otimes A^{\otimes 2}$ for matrix A given in (0.10). In the same spirit compute $A^{\otimes 4}$ using $A^{\otimes 3}$. Check your result by squaring $A^{\otimes 2}$.

2. Compute $A^{\otimes 2}$ for

$$A = \begin{pmatrix} 2 & 5 & -\infty \\ -\infty & 3 & 3 \\ 0 & -\infty & -\infty \end{pmatrix}.$$

3. In Section 0.4 the fifth train was added to the inner circuit. What would be the result if this fifth train were added to the outer circuit at S_2 instead? Write down the equations for that possibility and show that in that case the eigenvalue of the matrix concerned would still be equal to 4.

4. If you were on the board of the railway company, on which circuit would you add yet another train (the sixth one), whereby the fifth one remains on the inner circuit? The resulting eigenvalue (and interdeparture time) should be equal to 8/3. Can you relate this number to the now-existing bottleneck in the system?

5. Show that a coordinate transformation (in max-plus algebra, of course) between models (0.15) and (0.16) does not exist.

6. Check the derivation of equation (0.17), starting from the example in Section 0.1, and determine matrix B in (0.18). The latter matrix, of size 8×8, will contain quite a few elements $+\infty$.

7. Formulate the notion of eigenvalue and eigenvector for matrices in min-plus algebra. Without computation, what do you expect for the eigenvalue in min-algebra of matrix B of the above exercise?

8. Suppose that in the railway network of Figure 0.1 only three trains are available, one on each circuit. A possible interpretation of the inner circuit now is that it is a single track and that the direction in which the trains (actually the only train) use this track alternates. Derive a model of the form (0.9) for this layout and trains.

9. What is the minimal interdeparture time in the above exercise?

0.7 NOTES

The seminal work in max-plus algebra is [31]. Earlier traces, such as [30], exist. It was revitalized, in a system-theoretical setting, in [28] and subsequent papers of the "French school." The first railway application appeared in [15]. The book [5] describes the state of the art at the beginning of the 1990s. A more recent book with contributions by many authors is [48]. For more background information the reader is referred to the website www.maxplus.org.

The name *counter equation* is logical for equations of the form (0.18), since it refers to the question of how often an event has taken place up to a certain time. The generally accepted name *dater equation* for equations of the form (0.9) has French roots, due to the French school, which was seminal in the development of the theory. The word *date* in French refers to the calendar date, as it does in English, but in English there are other connotations. In historical perspective, *timer equation* might have been better to express the fact that the elements of the state x represent time instants.

PART I
Max-Plus Algebra

Chapter One

Max-Plus Algebra

In the previous chapter we described max-plus algebra in an informal way. The present chapter contains a more rigorous treatment of max-plus algebra. In Section 1.1 basic concepts are introduced, and algebraic properties of max-plus algebra are studied. Matrices and vectors over max-plus algebra are introduced in Section 1.2, and an important model, called *heap of pieces* or *heap model*, which can be described by means of max-plus algebra, is presented in Section 1.3. Finally, the projective space, a mathematical framework most convenient for studying limits, is introduced in Section 1.4.

1.1 BASIC CONCEPTS AND DEFINITIONS

Define $\varepsilon \stackrel{\text{def}}{=} -\infty$ and $e \stackrel{\text{def}}{=} 0$, and denote by $\mathbb{R}_{\max}$ the set $\mathbb{R} \cup \{\varepsilon\}$, where $\mathbb{R}$ is the set of real numbers. For elements $a, b \in \mathbb{R}_{\max}$, we define operations $\oplus$ and $\otimes$ by

$$a \oplus b \stackrel{\text{def}}{=} \max(a, b) \qquad \text{and} \qquad a \otimes b \stackrel{\text{def}}{=} a + b. \tag{1.1}$$

Clearly, $\max(a, -\infty) = \max(-\infty, a) = a$ and $a + (-\infty) = -\infty + a = -\infty$, for any $a \in \mathbb{R}_{\max}$, so that

$$a \oplus \varepsilon = \varepsilon \oplus a = a \qquad \text{and} \qquad a \otimes \varepsilon = \varepsilon \otimes a = \varepsilon, \tag{1.2}$$

for any $a \in \mathbb{R}_{\max}$. The above definitions are illustrated with some numerical examples as follows:

$$5 \oplus 3 = \max(5, 3) = 5,$$
$$5 \oplus \varepsilon = \max(5, -\infty) = 5,$$
$$5 \otimes \varepsilon = 5 - \infty = -\infty = \varepsilon,$$
$$e \oplus 3 = \max(0, 3) = 3,$$
$$5 \otimes 3 = 5 + 3 = 8.$$

The set $\mathbb{R}_{\max}$ together with the operations $\oplus$ and $\otimes$ is called *max-plus algebra* and is denoted by

$$\mathcal{R}_{\max} = (\mathbb{R}_{\max}, \oplus, \otimes, \varepsilon, e).$$

As in conventional algebra, we simplify the notation by letting the operation $\otimes$ have priority over the operation $\oplus$. For example,

$$5 \otimes -9 \oplus 7 \otimes 1$$

has to be understood as

$$(5 \otimes -9) \oplus (7 \otimes 1) .$$

Notice that $(5 \otimes -9) \oplus (7 \otimes 1) = 8$, whereas $5 \otimes (-9 \oplus 7) \otimes 1 = 13$.

The operations $\oplus$ and $\otimes$ defined in (1.1) have some interesting algebraic properties. For example, for $x, y, z \in \mathbb{R}_{\max}$, it holds that

$$
\begin{aligned}
x \otimes (y \oplus z) &= x + \max(y, z) \\
&= \max(x + y, x + z) \\
&= (x \otimes y) \oplus (x \otimes z) ,
\end{aligned}
$$

which in words means that $\otimes$ distributes over $\oplus$. Below we give a list of algebraic properties of max-plus algebra.

- Associativity:

$$\forall x, y, z \in \mathbb{R}_{\max} : \quad x \oplus (y \oplus z) = (x \oplus y) \oplus z$$

and

$$\forall x, y, z \in \mathbb{R}_{\max} : \quad x \otimes (y \otimes z) = (x \otimes y) \otimes z .$$

- Commutativity:

$$\forall x, y \in \mathbb{R}_{\max} : \quad x \oplus y = y \oplus x \qquad \text{and} \qquad x \otimes y = y \otimes x .$$

- Distributivity of $\otimes$ over $\oplus$:

$$\forall x, y, z \in \mathbb{R}_{\max} : \quad x \otimes (y \oplus z) = (x \otimes y) \oplus (x \otimes z) .$$

- Existence of a zero element:

$$\forall x \in \mathbb{R}_{\max} : \quad x \oplus \varepsilon = \varepsilon \oplus x = x .$$

- Existence of a unit element:

$$\forall x \in \mathbb{R}_{\max} : \quad x \otimes e = e \otimes x = x .$$

- The zero is absorbing for $\otimes$:

$$\forall x \in \mathbb{R}_{\max} : \quad x \otimes \varepsilon = \varepsilon \otimes x = \varepsilon .$$

- Idempotency of $\oplus$:

$$\forall x \in \mathbb{R}_{\max} : \quad x \oplus x = x .$$

Powers are introduced in max-plus algebra in the natural way using the associative property. We denote the set of natural numbers including zero by $\mathbb{N}$ and define for $x \in \mathbb{R}_{\max}$

$$x^{\otimes n} \overset{\text{def}}{=} \underbrace{x \otimes x \otimes \cdots \otimes x}_{n \text{ times}} \qquad (1.3)$$

for all $n \in \mathbb{N}$ with $n \neq 0$, and for $n = 0$ we define $x^{\otimes 0} \stackrel{\text{def}}{=} e \, (= 0)$. Observe that $x^{\otimes n}$, for any $n \in \mathbb{N}$, reads in conventional algebra as

$$x^{\otimes n} = \underbrace{x + x + \cdots + x}_{n \text{ times}} = n \times x \, .$$

For example,

$$5^{\otimes 3} = 3 \times 5 = 15.$$

Inspired by this we similarly introduce negative powers of real numbers, as in

$$8^{\otimes -2} = -2 \times 8 = -16 = 16^{\otimes -1} \, ,$$

for example. In the same vein, max-plus roots can be introduced as

$$x^{\otimes \alpha} = \alpha \times x \, ,$$

for $\alpha \in \mathbb{R}$. For example,

$$8^{\otimes \frac{1}{2}} = \frac{1}{2} \times 8 = 4$$

and

$$12^{\otimes -\frac{1}{4}} = -\frac{1}{4} \times 12 = -3 = 3^{\otimes -1} \, .$$

Continuing with the algebraic point of view, we show that max-plus algebra is an example of an algebraic structure, called a *semiring*, to be introduced next.

DEFINITION 1.1 *A semiring is a nonempty set R endowed with two binary operations $\oplus_R$ and $\otimes_R$ such that*

- *$\oplus_R$ is associative and commutative with zero element ε_R;*

- *$\otimes_R$ is associative, distributes over $\oplus_R$, and has unit element e_R;*

- *ε_R is absorbing for $\otimes_R$.*

Such a semiring is denoted by $\mathcal{R} = (R, \oplus_R, \otimes_R, \varepsilon_R, e_R)$. If $\otimes_R$ is commutative, then $\mathcal{R}$ is called commutative, *and if $\oplus_R$ is idempotent, then it is called* idempotent.

Max-plus algebra is an example of a commutative and idempotent semiring. Are there other meaningful semirings? The answer is yes, and a few examples are listed below.

Example 1.1.1

- *Identify $\oplus_R$ with conventional addition, denoted by $+$, and $\otimes_R$ with conventional multiplication, denoted by $\times$. Then the zero and unit element are $\varepsilon_R = 0$ and $e_R = 1$, respectively. The object $\mathcal{R}_{\text{st}} = (\mathbb{R}, +, \times, 0, 1)$ – the subscript st refers to "standard" – is an instance of a semiring over the real numbers. Since conventional multiplication is commutative, $\mathcal{R}_{\text{st}}$ is a commutative semiring. Note that $\mathcal{R}_{\text{st}}$ fails to be idempotent. However, as is well known, $\mathcal{R}_{\text{st}}$ is a ring and even a field with respect to the operations $+$ and $\times$. See the notes section for some further remarks on semirings and rings.*

- *Min-plus algebra is defined as* $\mathcal{R}_{\min} = (\mathbb{R}_{\min}, \oplus', \otimes, \varepsilon', e)$, *where* $\mathbb{R}_{\min} = \mathbb{R} \cup \{+\infty\}$, $\otimes'$ *is the operation defined by* $a \otimes' b \overset{\text{def}}{=} \min(a, b)$ *for all* $a, b \in \mathbb{R}_{\min}$, *and* $\varepsilon' \overset{\text{def}}{=} +\infty$. *Note that* $\mathcal{R}_{\min}$ *is an idempotent, commutative semiring.*

- *Consider* $\mathcal{R}_{\min,\max} = (\overline{\mathbb{R}}, \oplus', \oplus, \varepsilon', \varepsilon)$, *with* $\overline{\mathbb{R}} = \mathbb{R} \cup \{\varepsilon, \varepsilon'\}$, *and set* $\varepsilon \oplus \varepsilon' = \varepsilon' \oplus \varepsilon = \varepsilon'$. *Then* $\mathcal{R}_{\min,\max}$ *is an idempotent, commutative semiring. In the same vein,* $\mathcal{R}_{\max,\min} = (\overline{\mathbb{R}}, \oplus, \oplus', \varepsilon, \varepsilon')$ *is an idempotent, commutative semiring provided that one defines* $\varepsilon \oplus' \varepsilon' = \varepsilon' \oplus' \varepsilon = \varepsilon$.

- *As a last example of a semiring of a somewhat different nature, let* S *be a nonempty set. Denote the set of all subsets of* S *by* R; *then* $(R, \cup, \cap, \emptyset, S)$, *with* $\emptyset$ *the empty set, and* $\cup$ *and* $\cap$ *the set-theoretic union and intersection, respectively, is a commutative, idempotent semiring. The same applies to* $(R, \cap, \cup, S, \emptyset)$.

The above list of examples explains why we choose an algebraic approach. Any statement that is proved for a semiring will immediately hold in any of the above algebras. Apart from the structural insight this provides into the relationship between the different algebras, the algebraic approach also saves a lot of work.

To illustrate this, consider the following problem. Is it possible to define inverse elements (i.e., inverse with respect to the $\oplus_R$ operation) in an idempotent semiring? For example, consider an idempotent semiring $\mathcal{R} = (R, \oplus_R, \otimes_R, \varepsilon_R, e_R)$, with $\mathbb{R}$ included in R, such as the max-plus or min-plus semiring. For example, is it possible to find a solution of

$$5 \oplus_R x = 3 ? \tag{1.4}$$

As in conventional algebra, it is tempting to subtract 5 on both sides of the above equation in order to obtain

$$x = 3 \oplus_R (-5)$$

as a solution. However, is it possible to give meaning to -5 in the above equation? Take, for example, max-plus algebra. Then, equation (1.4) reads

$$\max(5, x) = 3. \tag{1.5}$$

Obviously, there exists no number that makes equation (1.5) true. On the other hand, in min-plus algebra, equation (1.5) reads

$$\min(5, x) = 3$$

and has the solution $x = 3$. Now interchange the numbers 3 and 5 in equation (1.4), yielding $3 \oplus_R x = 5$. This equation has no solution in min-plus algebra and has the obvious solution $x = 5$ in max-plus algebra.

Whether an equation has a solution may depend on the algebra. This raises the question whether a particular semiring (i.e., a particular interpretation of the symbols $\oplus_R$, $\otimes_R$, e_R, and ε_R) exists such that *all* equations of type (1.4) can be solved. The following lemma provides an answer.

Lemma 1.2 *Let $\mathcal{R} = (R, \oplus_R, \otimes_R, \varepsilon_R, e_R)$ be a semiring. Idempotency of $\oplus_R$ implies that inverse elements with respect to $\oplus_R$ do not exist.*

Proof. Suppose that $a \neq \varepsilon_R$ had an inverse element with respect to $\oplus_R$, say, b. In formula, this is

$$a \oplus_R b = \varepsilon_R.$$

Adding a on both sides of the above equation yields

$$a \oplus_R a \oplus_R b = a \oplus_R \varepsilon_R.$$

By idempotency, the left-hand side of the above equation equals $a \oplus_R b$, whereas the right-hand side is equal to a. Hence, we have

$$a \oplus_R b = a,$$

which contradicts $a \oplus_R b = \varepsilon_R$. $\square$

Lemma 1.2 thus gives a negative answer to the above question, because no idempotent semiring exists for which negative numbers can be defined. Observe that this does not contradict the fact that $\mathcal{R}_{\mathrm{st}}$, defined in Example 1.1.1, is a semiring because $\mathcal{R}_{\mathrm{st}}$ is not idempotent. The fact that we cannot subtract in an idempotent semiring explains why the methods encountered later, when studying max-plus algebra, will differ significantly from those in conventional algebra.

1.2 VECTORS AND MATRICES

In this section matrices over $\mathbb{R}_{\max}$ will be introduced. The set of $n \times m$ matrices with underlying max-plus algebra is denoted by $\mathbb{R}_{\max}^{n \times m}$. For $n \in \mathbb{N}$ with $n \neq 0$, define $\underline{n} \overset{\mathrm{def}}{=} \{1, 2, \ldots, n\}$. The element of a matrix $A \in \mathbb{R}_{\max}^{n \times m}$ in row i and column j is denoted by a_{ij}, for $i \in \underline{n}$ and $j \in \underline{m}$. Matrix A can then be written as

$$A = \begin{pmatrix} a_{11} & a_{12} & \cdots & a_{1m} \\ a_{21} & a_{22} & \cdots & a_{2m} \\ \vdots & \vdots & \ddots & \vdots \\ a_{n1} & a_{n2} & \cdots & a_{nm} \end{pmatrix}.$$

Occasionally, the element a_{ij} will also be denoted as

$$[A]_{ij}, \qquad i \in \underline{n}, j \in \underline{m}. \tag{1.6}$$

The sum of matrices $A, B \in \mathbb{R}_{\max}^{n \times m}$, denoted by $A \oplus B$, is defined by

$$[A \oplus B]_{ij} = a_{ij} \oplus b_{ij} \tag{1.7}$$
$$= \max\left(a_{ij}, b_{ij}\right),$$

for $i \in \underline{n}$ and $j \in \underline{m}$. For example, let

$$A = \begin{pmatrix} e & \varepsilon \\ 3 & 2 \end{pmatrix} \quad \text{and} \quad B = \begin{pmatrix} -1 & 11 \\ 1 & \varepsilon \end{pmatrix}; \tag{1.8}$$

then $[A \oplus B]_{11} = e \oplus -1 = \max(0, -1) = 0 = e$. Likewise, it follows that $[A \oplus B]_{12} = \varepsilon \oplus 11 = \max(-\infty, 11) = 11$, $[A \oplus B]_{21} = 3 \oplus 1 = \max(3, 1) = 3$, and $[A \oplus B]_{22} = 2 \oplus \varepsilon = \max(2, -\infty) = 2$. In matrix notation,

$$A \oplus B = \begin{pmatrix} e & 11 \\ 3 & 2 \end{pmatrix}.$$

Note that for $A, B \in \mathbb{R}_{\max}^{n \times m}$ it holds that $A \oplus B = B \oplus A$ (see Exercise 4).

For $A \in \mathbb{R}_{\max}^{n \times m}$ and $\alpha \in \mathbb{R}_{\max}$, the scalar multiple $\alpha \otimes A$ is defined by

$$[\alpha \otimes A]_{ij} = \alpha \otimes a_{ij} \tag{1.9}$$

for $i \in \underline{n}$ and $j \in \underline{m}$. For example, let A be defined as in (1.8) and take $\alpha = 2$; then $[2 \otimes A]_{11} = 2 \otimes e = 2 + 0 = 2$. Likewise, it follows that $[2 \otimes A]_{12} = \varepsilon$, $[2 \otimes A]_{21} = 5$, and $[2 \otimes A]_{22} = 4$, yielding, in matrix notation,

$$2 \otimes A = \begin{pmatrix} 2 & \varepsilon \\ 5 & 4 \end{pmatrix}.$$

For matrices $A \in \mathbb{R}_{\max}^{n \times l}$ and $B \in \mathbb{R}_{\max}^{l \times m}$, the matrix product $A \otimes B$ is defined as

$$[A \otimes B]_{ik} = \bigoplus_{j=1}^{l} a_{ij} \otimes b_{jk} \tag{1.10}$$

$$= \max_{j \in \underline{l}} \{a_{ij} + b_{jk}\}$$

for $i \in \underline{n}$ and $k \in \underline{m}$. This is just like in conventional algebra with $+$ replaced by max and $\times$ by $+$. Notice that $A \otimes B \in \mathbb{R}_{\max}^{n \times m}$, i.e., has n rows and m columns. For example, let A and B be defined as in (1.8); then the elements of $A \otimes B$ are given by

$$[A \otimes B]_{11} = e \otimes (-1) \oplus \varepsilon \otimes 1 = \max(0 - 1, -\infty + 1) = -1,$$
$$[A \otimes B]_{12} = e \otimes 11 \oplus \varepsilon \otimes \varepsilon = \max(0 + 11, -\infty - \infty) = 11,$$
$$[A \otimes B]_{21} = 3 \otimes (-1) \oplus 2 \otimes 1 = \max(3 - 1, 2 + 1) = 3,$$

and

$$[A \otimes B]_{22} = 3 \otimes 11 \oplus 2 \otimes \varepsilon = \max(3 + 11, 2 - \infty) = 14,$$

yielding, in matrix notation,

$$A \otimes B = \begin{pmatrix} -1 & 11 \\ 3 & 14 \end{pmatrix}.$$

Notice that the matrix product in general fails to be commutative. Indeed, for the above A and B

$$B \otimes A = \begin{pmatrix} 14 & 13 \\ 1 & \varepsilon \end{pmatrix} \neq A \otimes B.$$

Let $\mathcal{E}(n, m)$ denote the $n \times m$ matrix with all elements equal to ε, and denote by $E(n, m)$ the $n \times m$ matrix defined by

$$[E(n, m)]_{ij} \stackrel{\text{def}}{=} \begin{cases} e & \text{for } i = j, \\ \varepsilon & \text{otherwise.} \end{cases}$$

If $n = m$, then $E(n, n)$ is called the $n \times n$ identity matrix. When their dimensions are clear from the context, $\mathcal{E}(n, m)$ and $E(n, m)$ will also be written as $\mathcal{E}$ and E, respectively. It is easily checked (see exercise 5) that any matrix $A \in \mathbb{R}_{\max}^{n \times m}$ satisfies

$$A \oplus \mathcal{E}(n, m) = A = \mathcal{E}(n, m) \oplus A,$$

$$A \otimes E(m, m) = A = E(n, n) \otimes A.$$

Moreover, for $k \geq 1$ it holds that

$$A \otimes \mathcal{E}(m, k) = \mathcal{E}(n, k) \qquad \text{and} \qquad \mathcal{E}(k, n) \otimes A = \mathcal{E}(k, m).$$

For $\mathbb{R}_{\max}^{n \times m}$, the matrix addition $\oplus$, as defined in (1.7), is associative, commutative, and has zero element $\mathcal{E}(n, m)$. For $\mathbb{R}_{\max}^{n \times n}$ the matrix product $\otimes$, as defined in (1.10), is associative, distributive with respect to $\oplus$, has unit element $E(n, n)$, and $\mathcal{E}(n, n)$ is absorbing for $\otimes$.

The transpose of an element $A \in \mathbb{R}_{\max}^{n \times m}$, denoted by $A^\top$, is defined in the usual way by $[A^\top]_{ij} = a_{ji}$, for $i \in \underline{n}$ and $j \in \underline{m}$. As before, also in matrix addition and multiplication, the operation $\otimes$ has priority over the operation $\oplus$.

The elements of $\mathbb{R}_{\max}^n \overset{\text{def}}{=} \mathbb{R}_{\max}^{n \times 1}$ are called *vectors*. The jth element of a vector $x \in \mathbb{R}_{\max}^n$ is denoted by x_j, which, in the spirit of (1.6), also will be written as $[x]_j$. The vector in $\mathbb{R}_{\max}^n$ with all elements equal to e is called the *unit vector* and is denoted by $\mathbf{u}$; in formula, $[\mathbf{u}]_j = e$ for $j \in \underline{n}$. Notice that $\alpha \otimes \mathbf{u}$ denotes a vector with all elements equal to α, for any $\alpha \in \mathbb{R}_{\max}$. For any $j \in \underline{n}$, the jth column of the identity matrix $E(n, n)$ is called the jth *base vector* of $\mathbb{R}_{\max}^n$ and is denoted by e_j. Hence, the jth element of e_j has value e, while the other elements of e_j are equal to ε.

Note that for $A \in \mathbb{R}_{\max}^{n \times m}$ and $x \in \mathbb{R}_{\max}^m$, the product $A \otimes x$ is defined by (1.10) for $x = B$. Clearly, $A \otimes A$ and higher order powers of A are only defined for $A \in \mathbb{R}_{\max}^{n \times n}$, i.e., for matrices A that are square.

In the following a careful distinction will be made between $\mathbb{R}_{\max}^n$ (the set of n-dimensional vectors over $\mathbb{R}_{\max}$), $\mathbb{R}_{\max}^{n \times m}$ (the set of $n \times m$ matrices over $\mathbb{R}_{\max}$), and $\mathbb{R}_{\max}^{n \times n}$ (the set of square $n \times n$ matrices over $\mathbb{R}_{\max}$).

The structure

$$\mathcal{R}_{\max}^{n \times n} = \left(\mathbb{R}_{\max}^{n \times n}, \oplus, \otimes, \mathcal{E}, E \right),$$

with $\oplus$ and $\otimes$ as defined in (1.7) and (1.10), respectively, constitutes a noncommutative, idempotent semiring.

For $A \in \mathbb{R}_{\max}^{n \times n}$, denote the kth power of A by $A^{\otimes k}$ defined by

$$A^{\otimes k} \overset{\text{def}}{=} \underbrace{A \otimes A \otimes \cdots \otimes A}_{k \text{ times}}, \tag{1.11}$$

for $k \in \mathbb{N}$ with $k \neq 0$, and set $A^{\otimes 0} \overset{\text{def}}{=} E(n, n)$. The above definition is a straightforward extension of (1.3) to matrices. Notice that $[A^{\otimes k}]_{ij}$ has to be carefully distinguished from $(a_{ij})^{\otimes k}$. Indeed, the former is element (i, j) of the kth power of A, whereas the latter is the kth power of element (i, j) of A.

A mapping f from $\mathbb{R}^n_{\max}$ to $\mathbb{R}^n_{\max}$ is called *affine* if $f(x) = A \otimes x \oplus b$ for some $A \in \mathbb{R}^{n \times n}_{\max}$ and $b \in \mathbb{R}^n_{\max}$. If $b = \mathcal{E}$, then f is called *linear*. A recurrence relation $x(k+1) = f(x(k))$, for $k \in \mathbb{N}$, is called *affine* (resp., *linear*) if f is an affine (resp., linear) mapping.

A matrix $A \in \mathbb{R}^{n \times m}_{\max}$ is called *regular* if A contains at least one element different from ε in each row. Regularity is a mere technical condition, for if A fails to be regular, it contains redundant rows, and any system modeled by $x(k + 1) = A \otimes x(k)$ can also be modeled by considering a reduced regular version of A in which all redundant rows and related columns are skipped.

A matrix $A \in \mathbb{R}^{n \times n}_{\max}$ is called *strictly lower triangular* if $a_{ij} = \varepsilon$, for $1 \leq i \leq j \leq n$. If $a_{ij} = \varepsilon$, for $1 \leq i < j \leq n$, then A is called *lower triangular*. Matrix A is said to be *(strictly) upper triangular* if $A^\top$ is (strictly) lower triangular.

For countable sets the max operator has to be understood as a supremum. More formally, let $\{a_i : i \in \mathbb{N}\}$ be a countable set, with $a_i \in \mathbb{R}_{\max}$; then

$$\bigoplus_{i \geq 0} a_i \overset{\text{def}}{=} \bigoplus_{i=0}^{\infty} a_i \overset{\text{def}}{=} \sup_{i \geq 0} a_i.$$

For max-plus algebra one easily verifies Fubini's rule; namely, that for $\{a_{ij} \in \mathbb{R}_{\max} : i, j \in \mathbb{N}\}$,

$$\bigoplus_{i \geq 0} \bigoplus_{j \geq 0} a_{ij} = \bigoplus_{j \geq 0} \bigoplus_{i \geq 0} a_{ij}. \tag{1.12}$$

Indeed, for any $k, j \geq 0$ it follows that $a_{kj} \leq \bigoplus_{i \geq 0} a_{ij}$, implying that

$$\bigoplus_{j \geq 0} a_{kj} \leq \bigoplus_{j \geq 0} \bigoplus_{i \geq 0} a_{ij},$$

for any $k \geq 0$, and consequently that

$$\bigoplus_{k \geq 0} \bigoplus_{j \geq 0} a_{kj} \leq \bigoplus_{j \geq 0} \bigoplus_{i \geq 0} a_{ij}.$$

The inverse inequality follows from similar arguments.

1.3 A FIRST MAX-PLUS MODEL

In this section, we present an important example of a max-plus system, called a *heap model*. In a heap model, solid pieces are piled up according to a mechanism resembling the Tetris game. However, the pieces can only fall downwards vertically and cannot be moved horizontally or rotated. More specifically, consider the pieces labeled a, b, and c, as given in Figures 1.1 to 1.3. The pieces occupy columns out of a finite set of columns. The set of column numbers is given by $\mathcal{R}$, in the example the set $\{1, 2, \ldots, 5\}$. When the pieces are piled up according to a fixed sequence, like a b a c b, for example, this results in the heap shown in Figure 1.4.

Situations like the one pictured in Figure 1.4 typically arise in scheduling problems. Here, pieces represent tasks that compete for a limited number of resources,

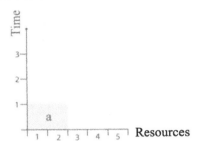

Figure 1.1: Piece a.

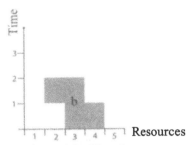

Figure 1.2: Piece b.

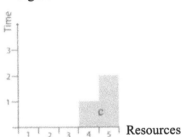

Figure 1.3: Piece c.

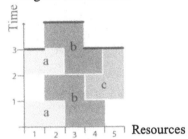

Figure 1.4: The heap w=abacb.

represented by the columns. The covering of a particular column by an individual piece can be interpreted as the amount of time required by the task (represented by the particular piece) of this resource.

Consider, for example, piece b in Figure 1.2. The idea is that piece b represents the time span for which resources have to be allocated in order to complete a certain task. More precisely, if processing the task is initiated at time t, then resource 2 will be occupied from time $t + 1$ until time $t + 2$, resource 3 will be occupied from time t until time $t + 2$, resource 4 will be occupied from time t until time $t + 1$, and resources 1 and 5 will not be occupied at all. The depiction of the processing times for this task given by piece b can be translated into mathematical terms by means of so-called contours. The upper and lower contours of a piece describe the covering of a piece lying on ground level. For example, the upper contour of piece b, denoted by $u(b)$, is

$$u(b) = (\varepsilon, 2, 2, 1, \varepsilon)^\top,$$

and the lower contour, denoted by $l(b)$, reads

$$l(b) = (\varepsilon, 1, e, e, \varepsilon)^\top,$$

where ε in the same location in $u(b)$ and $l(b)$ represents the fact that the piece does not cover the particular resource. The resources covered by piece b are denoted by $\mathcal{R}(b)$, so that $\mathcal{R}(b) = \{2, 3, 4\}$. For piece a it follows that $\mathcal{R}(a) = \{1, 2\}$. The upper contour of a is given by

$$u(a) = (1, 1, \varepsilon, \varepsilon, \varepsilon)^\top,$$

and the lower contour equals

$$l(a) = (e, e, \varepsilon, \varepsilon, \varepsilon)^\top.$$

As for piece c, it follows that $\mathcal{R}(c) = \{4, 5\}$,

$$u(c) = (\varepsilon, \varepsilon, \varepsilon, 1, 2)^\top,$$

and the lower contour equals

$$l(c) = (\varepsilon, \varepsilon, \varepsilon, e, e)^\top.$$

Before continuing, we will introduce some notation. Let $\mathcal{P}$ denote the finite set of pieces in the example $\mathcal{P} = \{a, b, c\}$. As already seen above a piece $\eta \in \mathcal{P}$ is characterized by its lower contour, denoted by $l(\eta)$, and its upper contour, denoted by $u(\eta)$. Moreover, the set of resources required by η is denoted by $\mathcal{R}(\eta)$. Let there be $n \in \mathbb{N}$, with $n \neq 0$, resources available. In our example we have $n = 5$. The upper and lower contours of a piece η are vectors over $\mathbb{R}^n_{\max}$, in formula $l(\eta), u(\eta) \in \mathbb{R}^n_{\max}$, such that

$$0 \leq l_r(\eta) \leq u_r(\eta) < \infty,$$

for $r \in \mathcal{R}(\eta)$, and

$$l_r(\eta) = u_r(\eta) = \varepsilon,$$

for $r \notin \mathcal{R}(\eta)$. Associate a matrix $M(\eta)$ with piece η through

$$[M(\eta)]_{rs} = \begin{cases} u_r(\eta) - l_s(\eta) & \text{for } r, s \in \mathcal{R}(\eta), \\ e & \text{for } s = r, \, r \notin \mathcal{R}(\eta), \\ \varepsilon & \text{otherwise.} \end{cases}$$

Elaborating on the upper and lower contours of pieces a, b, and c, respectively, the following matrices are obtained:

$$M(a) = \begin{pmatrix} 1 & 1 & \varepsilon & \varepsilon & \varepsilon \\ 1 & 1 & \varepsilon & \varepsilon & \varepsilon \\ \varepsilon & \varepsilon & e & \varepsilon & \varepsilon \\ \varepsilon & \varepsilon & \varepsilon & e & \varepsilon \\ \varepsilon & \varepsilon & \varepsilon & \varepsilon & e \end{pmatrix}, \quad M(b) = \begin{pmatrix} e & \varepsilon & \varepsilon & \varepsilon & \varepsilon \\ \varepsilon & 1 & 2 & 2 & \varepsilon \\ \varepsilon & 1 & 2 & 2 & \varepsilon \\ \varepsilon & e & 1 & 1 & \varepsilon \\ \varepsilon & \varepsilon & \varepsilon & \varepsilon & e \end{pmatrix},$$

and

$$M(c) = \begin{pmatrix} e & \varepsilon & \varepsilon & \varepsilon & \varepsilon \\ \varepsilon & e & \varepsilon & \varepsilon & \varepsilon \\ \varepsilon & \varepsilon & e & \varepsilon & \varepsilon \\ \varepsilon & \varepsilon & \varepsilon & 1 & 1 \\ \varepsilon & \varepsilon & \varepsilon & 2 & 2 \end{pmatrix}.$$

A sequence of pieces out of $\mathcal{P}$ is called a *heap*. For example, $w = \text{a b a c b}$ is a heap; see Figure 1.4. Denote the *upper contour* of heap w by a vector $x_{\mathcal{H}}(w) \in \mathbb{R}^n_{\max}$, where $(x_{\mathcal{H}}(w))_r$ is the height of the heap on column r; for example, $x_{\mathcal{H}}(\text{a b a c b}) = (3, 4, 4, 3, 3)^\top$, when starting from ground level. The upper contour of the heap a b a c b is indicated by the boldfaced line in Figure 1.4. For heap w and piece $\eta \in \mathcal{P}$, write $w\,\eta$ for the heap resulting from piling piece η on heap w. Note that the order in which the pieces fall is of importance. The upper contour follows the recurrence relation

$$[x_{\mathcal{H}}(w\,\eta)]_r = \max\left\{[M(\eta)]_{rs} + [x_{\mathcal{H}}(w)]_s : s \in \mathcal{R}\right\}, \qquad r \in \mathcal{R}, \qquad (1.13)$$

with initial upper contour $x_{\mathcal{H}}(\emptyset) = \mathbf{u}$, where $\emptyset$ denotes the empty heap. Elaborating on the notational power of the max-plus semiring, we can rewrite the above recurrence relation as

$$[x_{\mathcal{H}}(w\,\eta)]_r = \bigoplus_{s \in \mathcal{R}} [M(\eta)]_{rs} \otimes [x_{\mathcal{H}}(w)]_s, \qquad r \in \mathcal{R},$$

or, in a more concise way,

$$x_{\mathcal{H}}(w\,\eta) = M(\eta) \otimes x_{\mathcal{H}}(w).$$

In words, the upper contour of a heap of pieces follows a max-plus recurrence relation.

For a given sequence $\eta_1, \eta_2, \ldots, \eta_k$ of pieces, set, for notational convenience, $x_{\mathcal{H}}(k) = x_{\mathcal{H}}(\eta_1 \, \eta_2 \cdots \eta_k)$ and $M(k) = M(\eta_k)$. Then the upper contour follows the recurrence relation

$$x_{\mathcal{H}}(k + 1) = M(k + 1) \otimes x_{\mathcal{H}}(k), \qquad k \geq 1,$$

where $x_{\mathcal{H}}(0) = \mathbf{u}$.

In this context two kinds of limits are of interest, the first addressing the asymptotic growth rate of the heap and the second addressing the shape of the upper contour of the heap.

For a given sequence $\eta_k, k \in \mathbb{N}$, the *asymptotic growth rate* of the heap model is given by

$$\lim_{k \to \infty} \frac{1}{k} x_{\mathcal{H}}(k),$$

provided that the limit exists. For example, if $\eta_k, k \in \mathbb{N}$, represents a particular schedule, like $\eta_1 = \mathrm{a}, \eta_2 = \mathrm{b}, \eta_3 = \mathrm{c}, \eta_4 = \mathrm{a}, \eta_5 = \mathrm{b}, \eta_6 = \mathrm{c}$, and so forth, then the above limit measures the efficiency of schedule a b c.

For a given sequence $\eta_k, k \in \mathbb{N}$, the *asymptotic form of* $x_{\mathcal{H}}(k)$ can be studied, where *form* means the relative differences of the components of $x_{\mathcal{H}}(k)$. More precisely, in studying the shape of the upper contour the actual height of the heap is disregarded. To that end, the vector of relative differences in $x_{\mathcal{H}}(w)$, called the *shape vector*, is denoted by $s(w)$. For example, the shape of heap $w = \mathrm{a\,b\,a\,c\,b}$ in Figure 1.4 is obtained by letting the boldfaced line (the upper contour) sink to the ground level, yielding the vector $s(w) = (0, 1, 1, 0, 0)^\top$. More formally, the shape vector is defined as

$$s_r(w) = (x_{\mathcal{H}}(w))_r - \min\left\{(x_{\mathcal{H}}(w))_p : p \in \mathcal{R}\right\}, \qquad r \in \mathcal{R}.$$

Suppose that the sequence in which the pieces appear cannot be controlled (their arrivals may be triggered by an external source). For instance, $\eta_k, k \in \mathbb{N}$, is a random sequence such that piece a, b, and c appear with equal probability. Set $s(k) \stackrel{\text{def}}{=} s(\eta_1, \eta_2, \ldots, \eta_k)$. Since pieces fall in random order, $s(k)$ is a random variable. Using probabilistic arguments, one can identify sufficiency conditions such that the probability distribution of $s(k)$ converges to a limiting distribution, say, F. Hence, the asymptotic shape of the heap is given by the probability distribution F. By means of F, for example, the probability can be determined that the completion time of tasks typically differs more than t time units over the resources, yielding an indication on how well balanced the schedule η_k, $k \in \mathbb{N}$, is. See the notes section for references.

The asymptotic growth rate will be addressed in Section 3.2, and (the deterministic variant of) the limit of the shape vector will be addressed in Section 4.4.

1.4 THE PROJECTIVE SPACE

To sketch the idea of this section, let A be an $n \times n$ matrix with *positive* elements and let $x(k) \in \mathbb{R}^n$ be defined through
$$x(k+1) = A \otimes x(k), \qquad k \geq 0,$$
with $x(0) = x_0 \in \mathbb{R}^n$. Then, $x(k)$ is monotonically increasing, meaning that each of its components $x_i(k)$, $i \in \underline{n}$, is monotonically increasing. Taking the limit of $x(k)$ as k tends to ∞ will result in $(+\infty, \ldots, +\infty)^\top$ as the limiting vector. Indeed, revisit, for example, recurrence relation (0.10). The matrix describing the travel times has positive entries, and $x(k) = 4^{\otimes k} \otimes x_0$ for $x_0 = (1, 0)^\top$; see also Section 0.3. Notice that even though $x(k)$ diverges, the relative differences of $x(k)$ have a limit.

In this section, the modeling of differences within a vector will be explored more closely. Therefore, an equivalence relation on $\mathbb{R}^n_{\max}$ is introduced, denoted by $\cdot \parallel \cdot$, that is defined as
$$\forall y, z \in \mathbb{R}^n_{\max} : \quad y \parallel z \Leftrightarrow \exists \alpha \in \mathbb{R} : \quad y = \alpha \otimes z,$$
where the equation on the right-hand side should be read as $y_i = \alpha + z_i$ for all $i \in \underline{n}$. Two vectors $y, z \in \mathbb{R}^n_{\max}$ are said to be *colinear* (resp., *proportional*) if $y \parallel z$.

For $z \in \mathbb{R}^n_{\max}$, write $\bar{z}$ for the equivalence class $\{y \in \mathbb{R}^n_{\max} : y \parallel z\}$. Let $\mathbb{PR}^n_{\max}$ denote the *projective space*; that is, $\mathbb{PR}^n_{\max}$ is the quotient space of $\mathbb{R}^n_{\max}$ by the above equivalence relation. More formally,
$$\mathbb{PR}^n_{\max} = \{\bar{z} : z \in \mathbb{R}^n_{\max}\}.$$
The bar operator is the canonical projection of $\mathbb{R}^n_{\max}$ onto $\mathbb{PR}^n_{\max}$. In the same vein, denote by $\mathbb{PR}^n$ the quotient space of $\mathbb{R}^n$ by the above equivalence relation. With this terminology, the limit of the relative differences in the upper contour as k tends to ∞ now reads
$$\lim_{k \to \infty} \overline{x(k)}, \tag{1.14}$$
provided that the limit exists. For example, let v be an eigenvector of A and let λ be the corresponding eigenvalue. See, for instance, Section 0.3 for an introduction of these notions. Then, because all elements of A are positive, it can be shown (see Chapter 2) that v contains only finite elements and that $\lambda > 0$. Hence, $v \in \mathbb{R}^n$ and $\lambda > 0$ are such that $A \otimes v = \lambda \otimes v$. For $x(0) = v$, it then follows that $x(k) = \lambda^{\otimes k} \otimes v$ and $\overline{x(k)} = \bar{v}$. Hence, the shape is equal to $\bar{v}$ for any k. The projective space turns out to be a convenient mathematical space for speaking about limits of sequences $\{x(k) : k \in \mathbb{N}\}$ stemming from max-plus recurrence relations.

Notice that it actually has not been explained what it means when the above limit is said to exist. A precise definition will be provided in Section 4.4.

1.5 EXERCISES

1. Show that the algebraic structures defined in Example 1.1.1 are indeed semirings.

2. Compute the following:

    (a) $-8 \otimes \varepsilon$

    (b) $(-1)^{\otimes \frac{1}{2}} \left(\stackrel{\text{def}^{\otimes}}{=} \sqrt{-1} \right)$

    (c) the product of the next two matrices

$$\begin{pmatrix} 1 & \varepsilon \\ \varepsilon & e \\ 2 & 1 \\ \varepsilon & \varepsilon \end{pmatrix}, \quad \begin{pmatrix} 4 & 2 & \varepsilon & 8 & e \\ -1 & \varepsilon & \varepsilon & 7 & 1 \end{pmatrix}$$

3. Show that for any $n \in \mathbb{N}$ numbers $x, y, z \in \mathbb{R}_{\max}$ exist such that

$$z^{\otimes n} = x^{\otimes n} \oplus y^{\otimes n},$$

   i.e., Fermat's theorem is not true over $\mathbb{R}_{\max}$.

4. Show that for $A, B, C \in \mathbb{R}_{\max}^{n \times n}$ the following properties are true:

    (a) Associativity: $A \oplus (B \oplus C) = (A \oplus B) \oplus C$ and $A \otimes (B \otimes C) = (A \otimes B) \otimes C$

    (b) Commutativity: $A \oplus B = B \oplus A$

    (c) Distributivity of $\otimes$ over $\oplus$: $A \otimes (B \oplus C) = (A \otimes B) \oplus (A \otimes C)$

5. Let $A \in \mathbb{R}_{\max}^{n \times m}$. Show that

$$A \oplus \mathcal{E}(n, m) = A = \mathcal{E}(n, m) \oplus A,$$

$$A \otimes \mathcal{E}(m, k) = \mathcal{E}(n, k),$$

   and

$$\mathcal{E}(k, n) \otimes A = \mathcal{E}(k, m),$$

   for $k \geq 1$. Moreover, show

$$A \otimes E(m, m) = A = E(n, n) \otimes A.$$

6. (a) Show that for $\mathcal{R}_{\min,\max}$ to be a semiring, one needs to define $\max(+\infty, -\infty) = \max(-\infty, +\infty) = +\infty$.

    (b) Show that for $\mathcal{R}_{\max,\min}$ to be a semiring, one needs to define $\min(+\infty, -\infty) = \min(-\infty, +\infty) = -\infty$.

    (c) Show that an expression in terms of $\otimes_R$ and $\oplus_R$, in general, will attain different numerical values when evaluated in $\mathcal{R}_{\min,\max}$ or in $\mathcal{R}_{\max,\min}$.

7. Show that (1.13) is indeed the correct recurrence relation.

8. Let $x, y \in \mathbb{R}_{\max}^n$ be such that $\alpha \otimes x = y$ for some $\alpha \in \mathbb{R}$. Show that $\bar{x} = \bar{y}$.

9. Show that for $\bar{x}, \bar{y} \in \mathbb{PR}_{\max}^n$ it generally does not hold that $\bar{x} \oplus \bar{y} = \overline{x \oplus y}$.

10. A semiring $\mathcal{R}$ is said to have zero-divisors if elements $x, y \neq \varepsilon_R$ exist such that $x \otimes_R y = \varepsilon_R$. Show that $\mathcal{R}_{\max}$ is zero-divisor free and that, for $n > 1$, $\mathcal{R}_{\max}^{n \times n}$ possesses zero-divisors. (Hint: Use matrices

$$A = \begin{pmatrix} \varepsilon & \varepsilon \\ \varepsilon & e \end{pmatrix}, \quad B = \begin{pmatrix} e & \varepsilon \\ \varepsilon & \varepsilon \end{pmatrix},$$

and show that $A \otimes B = \mathcal{E}$.)

11. Let $\mathcal{B} = \{\varepsilon, e\}$. Then $(\mathcal{B}, \oplus, \otimes, \varepsilon, e)$ is called Boolean algebra. Show that Boolean algebra is a semiring.

1.6 NOTES

For an extensive discussion of max-plus algebra and similar structures we refer to [5]. An early reference is [31]. A historical overview of the beginnings of max-plus theory can be found in [36]. This article also contains many more examples of semirings. For more details on idempotency, see [48]. In [21] the solvability of sets of equations over $\mathcal{R}_{\min,\max}$ and $\mathcal{R}_{\max,\min}$ is treated.

From the semiring theory point of view, it seems more natural to use the symbols $+$ and $\times$ instead of $\oplus$ and $\otimes$ and, for consistency, 0 and 1 for the zero and the unit. However, in applications, often hybrid formulas are encountered containing conventional addition and multiplication as well as addition and multiplication in a semiring. For this reason, the notation for semirings will be carefully distinguished from that for operations in conventional algebra.

The term *semiring* originates from the fact that $(R, \oplus_R, \varepsilon_R)$ in the definition of a semiring is a semigroup. Indeed, since that inverse elements with respect to $\oplus_R$ do not exist, it follows that $(R, \oplus_R, \varepsilon_R)$ is not a group but a semigroup (and even a monoid). Consequently, $(R, \oplus_R, \otimes_R, \varepsilon_R, e_R)$, with all the properties stated in Definition 1.1, is not a ring but is just a semiring. In literature, idempotent semirings are also called *dioids*; see [5]. Observe that $\mathcal{R}_{\max}$ is by no means an algebra in the classical sense. The name *max-plus algebra* is only historically justified, and the correct name for $\mathcal{R}_{\max}$ would be *idempotent semiring* or *dioid* (which might explain why the name *max-plus algebra* is still predominant in the literature). The book [45] discusses general aspects of idempotent structures, also in the infinite-dimensional case (in connection with the Hamilton-Jacobi equation). A reference book on general algebraic structures is [44]. The books [64] and [57] focus on applications in physics.

Heap models were introduced in [40] and further studied in [42]; see also the references therein for more details. For applications of heap models to scheduling we refer to [19], [41], and [42]. A variant of the heap model is to consider colored pieces. The basic idea is to normalize the heap, consisting of differently colored pieces, to a certain fixed height. When piling up pieces, the overall height of the heap does not change but its average color does. For example, having only two colors, say, red and blue, the heap will in the limit attain a certain shade of purple representing the limit regime of the schedule.

In discrete-time optimal control or, in Markovian decision theory one encounters the equation

$$V(k, x) = \max_u \big(V(k + 1, f(x, u)) + g(x, u) \big),$$

which is a consequence of Bellman's principle of optimality. The underlying model is $x(k + 1) = f(x(k), u(k))$ and the costs during time step k are $g(x(k), u(k))$. The function V is the value function. This equation, with the operations addition and maximization can be interpreted and analyzed in the sense of max-plus algebra; see [1] and [65].

Chapter Two

Spectral Theory

This chapter is devoted to spectral theory of matrices over the max-plus semiring. In Section 2.1 we will study the relation between graphs and matrices over the max-plus semiring. The basic observation is that any square matrix can be translated into a weighted graph (to be defined shortly) and that products and powers of matrices over the max-plus semiring have entries with a nice graph-theoretical interpretation. This interpretation will be further studied in Section 2.2. The key result will be that, under mild conditions, a square matrix over the max-plus semiring possesses a unique eigenvalue that equals the maximal average weight of circuits in the associated graph. This result plays a crucial role in numerical analysis of systems that can be modeled by max-plus recurrence relations. The chapter is concluded with Section 2.3 in which some elementary results on certain max-plus linear equations are presented.

2.1 MATRICES AND GRAPHS

A *directed graph* $\mathcal{G}$ is a pair $(\mathcal{N}, \mathcal{D})$, where $\mathcal{N}$ is a finite set of elements called *nodes* (or *vertices*) and $\mathcal{D} \subset \mathcal{N} \times \mathcal{N}$ is a set of ordered pairs of nodes called *arcs* (or *edges*). The word *ordered* means that the arcs (i, j) and (j, i) will be distinguished. If $(i, j) \in \mathcal{D}$, then we say that $\mathcal{G}$ contains an arc from i to j, and the arc (i, j) is called an *incoming arc* at j and an *outgoing arc* at i. Suppose that $(i, j) \in \mathcal{D}$, but $(j, i) \notin \mathcal{D}$; then an arc from i to j exists but there isn't an arc from j to i. This distinction in the direction of an arc explains the name *directed graph*. A directed graph is also called a *digraph* in the literature. A directed graph is called *weighted* if a weight $w(i, j) \in \mathbb{R}$ is associated with any arc $(i, j) \in \mathcal{D}$. From now on we will deal exclusively with weighted directed graphs and will refer to them as "graphs" for simplicity.

To any $n \times n$ matrix A over $\mathbb{R}_{\max}$ a graph can be associated, called the *communication graph* of A. The graph will be denoted by $\mathcal{G}(A)$. The set of nodes of the graph is given by $\mathcal{N}(A) = \underline{n}$, and a pair $(i, j) \in \underline{n} \times \underline{n}$ is an arc of the graph if $a_{ji} \neq \varepsilon$ (this is not a typo!), i.e., in symbols $(i, j) \in \mathcal{D}(A) \Leftrightarrow a_{ji} \neq \varepsilon$, where $\mathcal{D}(A)$ denotes the set of arcs of the graph.

For any two nodes i, j, a sequence of arcs $p = ((i_k, j_k) \in \mathcal{D}(A) : k \in \underline{m})$, such that $i = i_1$, $j_k = i_{k+1}$, for $k < m$, and $j_m = j$ is called a *path* from i to j. The path is then said to consist of the nodes $i = i_1, i_2, \ldots, i_m, j_m = j$ and to have *length* m. The latter will be denoted as $|p|_l = m$. Further, if $i = j$, then the path is called a *circuit*. A circuit $p = ((i_1, i_2), (i_2, i_3), \ldots, (i_m, i_1))$ is called *elementary* if, restricted to the circuit, each of its nodes has only one incoming and one outgoing arc or, more formally, if nodes i_k and i_l are different for $k \neq l$. A circuit consisting of just one arc, from a node to itself, is also called a *self-loop*.

The set of all paths from i to j of length $m \geq 1$ is denoted by $P(i, j; m)$. For an arc (i, j) in $\mathcal{G}(A)$, the weight of (i, j) is given by a_{ji} (again, this is not a typo!), and the *weight* of a path in $\mathcal{G}(A)$ is defined by the sum of the weights of all arcs constituting the path. More formally, for $p = ((i_1, i_2), (i_2, i_3), \ldots, (i_m, i_{m+1})) \in P(i, j; m)$ with $i = i_1$ and $j = i_{m+1}$, define the weight of p, denoted by $|p|_w$, through

$$|p|_w = \bigotimes_{k=1}^{m} a_{i_{k+1} i_k}.$$

Note that in conventional notation $|p|_w = \sum_{k=1}^{m} a_{i_{k+1} i_k}$. The *average weight* of a path p is given by $|p|_w / |p|_1$. For circuits the notions of weight, length, and average weight are defined similarly as for paths. Also, the phrase *circuit mean* is used instead of the phrase *average circuit weight*.

Paths in $\mathcal{G}(A)$ can be combined in order to construct a new path. For example, let $p = ((i_1, i_2), (i_2, i_3))$ and $q = ((i_3, i_4), (i_4, i_5))$ be two paths in $\mathcal{G}(A)$. Then,

$$p \circ q = ((i_1, i_2), (i_2, i_3), (i_3, i_4), (i_4, i_5))$$

is a path in $\mathcal{G}(A)$ as well. The operation $\circ$ is called the *concatenation of paths*. Clearly, the operation is not commutative, even when both $p \circ q$ and $q \circ p$ are defined.

Example 2.1.1   *Let*

$$A = \begin{pmatrix} \varepsilon & 15 & \varepsilon \\ \varepsilon & \varepsilon & 14 \\ 10 & \varepsilon & 12 \end{pmatrix}.$$

The communication graph of A is shown in Figure 2.1. The graph $\mathcal{G}(A)$ has node

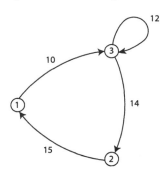

Figure 2.1: The communication graph of matrix A in Example 2.1.1.

set $\mathcal{N}(A) = \{1, 2, 3\}$ and arc set $\mathcal{D}(A) = \{(1, 3), (3, 2), (2, 1), (3, 3)\}$. Specifically, $\mathcal{G}(A)$ consists of two elementary circuits, namely, $\rho = ((1, 3), (3, 2), (2, 1))$ and $\theta = (3, 3)$. The weight of ρ is given by

$$|\rho|_w = a_{12} + a_{23} + a_{31} = 39,$$

and the length of ρ equals $|\rho|_1 = 3$. Circuit θ has weight $|\theta|_w = a_{33} = 12$ and is of length 1.

The communication graph $\mathcal{G}(A)$ and powers of A are closely related to each other. As will be proved in the next theorem, the element $[A^{\otimes k}]_{ji}$ yields the maximal weight of a path of length k from node i to node j, provided that such a path exists.

THEOREM 2.1 *Let $A \in \mathbb{R}_{max}^{n \times n}$. It holds for all $k \geq 1$ that*

$$\left[A^{\otimes k}\right]_{ji} = \max\left\{\, |p|_w \, : \, p \in P(i, j; k) \,\right\},$$

where $[A^{\otimes k}]_{ji} = \varepsilon$ in the case where $P(i, j; k)$ is empty, i.e., when no path of length k from i to j exists in $\mathcal{G}(A)$.

Proof. The proof is done by induction. Let (i, j) be an arbitrary element of $\underline{n} \times \underline{n}$. For $k = 1$, paths in $P(i, j; k)$ consist only of one arc whose weight is by definition $[A]_{ji}$. If $[A]_{ji} = \varepsilon$, then there is no arc (i, j) in $\mathcal{G}(A)$ and $P(i, j; 1) = \emptyset$.

Suppose now that the theorem holds true for k. Consider $p \in P(i, j; k + 1)$ and assume that there exists at least one path in $P(i, j; k + 1)$. Such a path, say, p can be split up into a subpath of length k running from i to some node l, and a path consisting of one arc from l to j or, more formally,

$$p = \hat{p} \circ (l, j) \quad \text{with} \quad \hat{p} \in P(i, l; k).$$

The maximal weight of any path in $P(i, j; k + 1)$ can thus be obtained from

$$\max_{l \in \underline{n}} \left([A]_{jl} + \max\left\{ |\hat{p}|_w \, : \, \hat{p} \in P(i, l; k) \right\} \right). \tag{2.1}$$

In accordance with the induction hypothesis, it holds that

$$\max\left\{ |\hat{p}|_w \, : \, \hat{p} \in P(i, l; k) \right\} = \left[A^{\otimes k}\right]_{li},$$

and the expression for the maximal weight of a path from i to j of length $(k + 1)$ in (2.1) reads

$$\max_{l \in \underline{n}} \left(a_{jl} + \left[A^{\otimes k}\right]_{li} \right) = \bigoplus_{l=1}^{n} a_{jl} \otimes \left[A^{\otimes k}\right]_{li}$$

$$= \left[A \otimes A^{\otimes k}\right]_{ji} = \left[A^{\otimes(k+1)}\right]_{ji}.$$

Now turn to the case in which $P(i, j; k + 1) = \emptyset$; i.e., there exists no path of length $k + 1$ from i to j. Obviously, this implies that for any node l it holds true that either there exists no path of length k from i to l or there exists no arc from l to j (or both). Hence, $P(i, j; k + 1) = \emptyset$ implies that for any l at least one of the values a_{jl} and $\left[A^{\otimes k}\right]_{li}$ equals ε. As a consequence,

$$\left[A^{\otimes(k+1)}\right]_{ji} = \varepsilon,$$

which completes the proof of the theorem. $\qquad\qquad\square$

The above theorem is illustrated with the following example.

Example 2.1.2   *Consider a railway network with a set of stations $\underline{n}$, with $n \geq 1$. The railway system can be mapped on a graph $\mathcal{G} = (\mathcal{N}, \mathcal{D})$ in the following way. Set $\mathcal{N} = \underline{n}$, and for $i, j \in \underline{n}$, let (i, j) be an arc in $\mathcal{D}$ if there is a direct connection from station i to station j in the railway network. To arc $(i, j) \in \mathcal{D}$, associate weight a_{ji}, where a_{ji} denotes the travel time from i to j. Suppose one is interested in the maximal time to travel from a particular station i to a particular station j in m steps. A direct way of computing this number would be to check all paths of length m from i to j in $\mathcal{G}$ and to compare their weights. An alternative way for computing the maximal travel time is as follows. Let A be the $n \times n$ matrix whose elements represent the weights of arcs in $\mathcal{G}$, where $a_{ji} = \varepsilon$ if $(i, j) \notin \mathcal{D}$. Then the maximal weight of a path from i to j of length m is given by $[A^{\otimes m}]_{ji}$; i.e., the maximal weight is given by the element (j, i) of the mth power of A.*

For $A \in \mathbb{R}_{\max}^{n \times n}$, let

$$A^+ \stackrel{\text{def}}{=} \bigoplus_{k=1}^{\infty} A^{\otimes k}. \tag{2.2}$$

The element $[A^+]_{ij}$ yields the maximal weight of any path from j to i (the value $[A^+]_{ij} = +\infty$ is possible). Indeed, by definition

$$[A^+]_{ij} = \max\left\{ [A^{\otimes k}]_{ij} : k \geq 1 \right\},$$

where $[A^{\otimes k}]_{ij}$ is the maximal weight of a path from j to i of length k; see Theorem 2.1.

LEMMA 2.2 *Let $A \in \mathbb{R}_{\max}^{n \times n}$ be such that any circuit in $\mathcal{G}(A)$ has average circuit weight less than or equal to e. Then, it holds that*

$$A^+ = \bigoplus_{k=1}^{\infty} A^{\otimes k} = A \oplus A^{\otimes 2} \oplus A^{\otimes 3} \oplus \cdots \oplus A^{\otimes n} \in \mathbb{R}_{\max}^{n \times n}.$$

Proof. Since A is of dimension n, all paths in $\mathcal{G}(A)$ from i to j of length greater than n are necessarily made up of at least one circuit and a path from i to j of length at most n. Because circuits in $\mathcal{G}(A)$ have nonpositive weights, it follows that

$$[A^+]_{ji} \leq \max\{[A^{\otimes k}]_{ji} : k \in \underline{n}\},$$

which concludes the proof of the lemma. $\square$

We conclude this section with a discussion of graph-theoretical concepts that we will need later on. Let $\mathcal{G} = (\mathcal{N}, \mathcal{D})$ denote a graph with node set $\mathcal{N}$ and arc set $\mathcal{D}$. For $i, j \in \mathcal{N}$, node j is said to be reachable from node i, denoted as $i \mathcal{R} j$, if there exists a path from i to j. A graph $\mathcal{G}$ is called *strongly connected* if for any two nodes $i, j \in \mathcal{N}$, node j is reachable from node i. A matrix $A \in \mathbb{R}_{\max}^{n \times n}$ is called *irreducible* if its communication graph $\mathcal{G}(A)$ is strongly connected. If a matrix is not irreducible, it is called *reducible*.

To better deal with graphs that are not strongly connected, we say for nodes $i, j \in \mathcal{N}$ that node j *communicates with* node i, denoted as $i \mathcal{C} j$, if either $i = j$ or there exists a path from i to j and a path from j to i. Hence, $i \mathcal{C} j \iff i = j$ or $[i \mathcal{R} j$

and $j\mathcal{R}i$]. Note that the relation "communicates with" is an equivalence relation. Indeed, its reflexivity and symmetry follow by definition, and its transitivity follows by the concatenation of paths.

If a graph $\mathcal{G} = (\mathcal{N}, \mathcal{D})$ is not strongly connected, then not all nodes of $\mathcal{N}$ communicate with each other. In this case, given a node, say, node i, it is possible to distinguish the subset of nodes that communicate with i from the subset of nodes that do not communicate with i. In the first subset all nodes communicate with each other, whereas in the second subset not all nodes necessarily communicate with each other. In the latter case a further subdivision of the nodes is possible. Repeated application of the previous idea therefore yields that the node set $\mathcal{N}$ can be partitioned as $\mathcal{N}_1 \cup \mathcal{N}_2 \cup \cdots \cup \mathcal{N}_q$, where $\mathcal{N}_r$, $r \in \underline{q}$, denotes a subset of nodes that communicate with each other but not with other nodes of $\mathcal{N}$. Recall that a partitioning of a set is a division into nonempty subsets such that the joint union is the whole set and the mutual intersections are all empty.

Given the above partitioning of $\mathcal{N}$, it is possible to focus on subgraphs of $\mathcal{G}$, denoted by $\mathcal{G}_r = (\mathcal{N}_r, \mathcal{D}_r)$, $r \in \underline{q}$, where $\mathcal{D}_r$ denotes the subset of $\mathcal{D}$ of arcs that have both the begin node and the end node in $\mathcal{N}_r$. If $\mathcal{D}_r \neq \emptyset$, the subgraph $\mathcal{G}_r = (\mathcal{N}_r, \mathcal{D}_r)$ is known as a maximal strongly connected subgraph (m.s.c.s.) of $\mathcal{G} = (\mathcal{N}, \mathcal{D})$. By definition, nodes in $\mathcal{N}_r$ do not communicate with nodes outside $\mathcal{N}_r$. However, it can happen that $i\mathcal{R}j$ for some $i \in \mathcal{N}_r$ and $j \in \mathcal{N}_{r'}$ with $r \neq r'$, but then the converse (i.e., $j\mathcal{R}i$) does not hold. We denote by $[i] \overset{\text{def}}{=} \{j \in \mathcal{N} : i\mathcal{C}j\}$ the set of nodes containing node i that communicate with each other. These nodes together form the equivalence class in which i is contained. Hence, given node $i \in \mathcal{N}$, there exists an $r \in \underline{q}$ such that $i \in \mathcal{N}_r$ and $[i] = \mathcal{N}_r$.

Note that the above partitioning covers *all* nodes of $\mathcal{N}$. If a node of $\mathcal{G}$ is contained in one or more circuits, it communicates with certain other nodes or with itself in case one of the circuits actually is a self-loop. In any case, the arc set of the associated subgraph is not empty. However, if the graph $\mathcal{G}$ contains a node that is not contained in any circuit of $\mathcal{G}$, say, node i, then node i does not communicate with other nodes and it communicates only with itself. Then, by definition, node i forms an equivalence class on its own, so that $[i] = \{i\}$. Because there does not even exist an arc from i to itself, it follows that the associated subgraph is given by $([i], \emptyset)$; i.e., the node set consists of node i only and the arc set is empty. Further, although it is not strongly connected, $([i], \emptyset)$ will be referred to as an m.s.c.s. This is merely done for convenience. Hence, in the following all subgraphs $\mathcal{G}_r = (\mathcal{N}_r, \mathcal{D}_r)$, $r \in \underline{q}$, introduced above are referred to as m.s.c.s.'s.

We define the *reduced graph*, denoted by $\widetilde{\mathcal{G}} = (\widetilde{\mathcal{N}}, \widetilde{\mathcal{D}})$, by $\widetilde{\mathcal{N}} = \{[i_1], \ldots, [i_q]\}$ and $([i_r], [i_s]) \in \widetilde{\mathcal{D}}$ if $r \neq s$ and there exists an arc $(k, l) \in \mathcal{D}$ for some $k \in [i_r]$ and $l \in [i_s]$. Hence, the number of nodes in the reduced graph is exactly the number of m.s.c.s.'s in the graph. The reduced graph models the interdependency of m.s.c.s.'s.

Note that the reduced graph does not contain circuits. Indeed, if the reduced graph would contain a circuit, then two or more m.s.c.s.'s would be connected to each other by means of a circuit, forming a new m.s.c.s. larger than the m.s.c.s.'s it contains. However, this would contradict the fact that these subgraphs already were maximal and strongly connected.

Let A_{rr} denote the matrix obtained by restricting A to the nodes in $[i_r]$, for all $r \in \underline{q}$, i.e., $[A_{rr}]_{kl} = a_{kl}$ for all $k, l \in [i_r]$. Notice that for all $r \in \underline{q}$ either A_{rr} is irreducible or $A_{rr} = \varepsilon$. It is easy to see that because the reduced graph does not contain any circuits, the original reducible matrix A, possibly after a relabeling of the nodes in $\mathcal{G}(A)$, can be written in the form

$$
\begin{pmatrix}
A_{11} & A_{12} & \cdots & \cdots & A_{1q} \\
\varepsilon & A_{22} & \cdots & \cdots & A_{2q} \\
\varepsilon & \varepsilon & A_{33} & & \vdots \\
\vdots & \vdots & \ddots & \ddots & \vdots \\
\varepsilon & \varepsilon & \cdots & \varepsilon & A_{qq}
\end{pmatrix},
$$

with matrices $A_{sr}, 1 \leq s < r \leq q$, of appropriate size. Each finite entry in A_{sr} corresponds to an arc from a node in $[i_r]$ to a node in $[i_s]$. The block upper triangular form shown above is said to be a *normal form* of matrix A. Note that the normal form of a matrix is not unique.

The notion of the cyclicity of a graph plays an important role in this book.

DEFINITION 2.3 *The cyclicity of a graph $\mathcal{G}$, denoted by $\sigma_{\mathcal{G}}$, is defined as follows:*

- *If $\mathcal{G}$ is strongly connected, then its cyclicity equals the greatest common divisor of the lengths of all elementary circuits in $\mathcal{G}$. If $\mathcal{G}$ consists of just one node without a self-loop, then its cyclicity is defined to be one.*

- *If $\mathcal{G}$ is not strongly connected, then its cyclicity equals the least common multiple of the cyclicities of all maximal strongly connected subgraphs of $\mathcal{G}$.*

We continue with further definitions of some graph-theoretical notions. The set of *direct predecessors* of node i is denoted by $\pi(i)$; more formally,

$$
\pi(i) \stackrel{\text{def}}{=} \{ j \in \underline{n} : (j, i) \in \mathcal{D} \}.
$$

Moreover, denote the set of all *predecessors* of node i by

$$
\pi^+(i) \stackrel{\text{def}}{=} \{ j \in \underline{n} : j\mathcal{R}i \},
$$

and set $\pi^*(i) = \{i\} \cup \pi^+(i)$. In words, $\pi(i)$ is the set of nodes immediately upstream of i; $\pi^+(i)$ is the set of all nodes from which node i can be reached; and $\pi^*(i)$ is the set of all nodes from which node i can be reached, including node i itself. In the same vein, denote the set of *direct successors* of node i by $\sigma(i)$, more formally,

$$
\sigma(i) \stackrel{\text{def}}{=} \{ j \in \underline{n} : (i, j) \in \mathcal{D} \};
$$

write

$$
\sigma^+(i) \stackrel{\text{def}}{=} \{ j \in \underline{n} : i\mathcal{R}j \}
$$

for the set of all *successors* of node i; and set $\sigma^*(i) = \{i\} \cup \sigma^+(i)$. In words, $\sigma(i)$ is the set of nodes immediately downstream of i; $\sigma^+(i)$ is the set of all nodes that can be reached from node i; and $\sigma^*(i)$ is the set of all nodes that can be reached from i, including node i itself.

Example 2.1.3 *Let*

$$A = \begin{pmatrix} \varepsilon & 0 & \varepsilon & \varepsilon & \varepsilon & \varepsilon & \varepsilon & \varepsilon & \varepsilon & \varepsilon \\ \varepsilon & \varepsilon & -3 & \varepsilon & \varepsilon & \varepsilon & \varepsilon & \varepsilon & \varepsilon & \varepsilon \\ \varepsilon & 4 & \varepsilon & 0 & \varepsilon & \varepsilon & \varepsilon & \varepsilon & \varepsilon & \varepsilon \\ 0 & \varepsilon & \varepsilon & \varepsilon & \varepsilon & \varepsilon & \varepsilon & \varepsilon & \varepsilon & \varepsilon \\ \varepsilon & 16 & \varepsilon & \varepsilon & \varepsilon & -5 & \varepsilon & \varepsilon & \varepsilon & \varepsilon \\ \varepsilon & \varepsilon & \varepsilon & \varepsilon & \varepsilon & \varepsilon & 0 & \varepsilon & \varepsilon & \varepsilon \\ \varepsilon & \varepsilon & \varepsilon & \varepsilon & 9 & \varepsilon & \varepsilon & \varepsilon & \varepsilon & \varepsilon \\ \varepsilon & \varepsilon & \varepsilon & \varepsilon & \varepsilon & \varepsilon & \frac{1}{2} & \varepsilon & \varepsilon & \varepsilon \\ \varepsilon & \varepsilon & \varepsilon & \varepsilon & \varepsilon & 6 & \varepsilon & \varepsilon & \varepsilon & \varepsilon \\ 9 & \varepsilon & \varepsilon & \varepsilon & \varepsilon & \varepsilon & \varepsilon & \varepsilon & \varepsilon & \varepsilon \end{pmatrix}.$$

The communication graph of A is shown in Figure 2.2. The graph $\mathcal{G}(A)$ has node

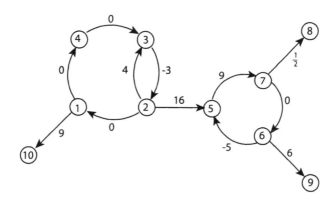

Figure 2.2: The communication graph of matrix A in Example 2.1.3.

set $\mathcal{N}(A) = \{1, 2, 3, \ldots, 10\}$ and arc set $\mathcal{D}(A) = \{(2,1), (3,2), (2,3), (4,3), (1,4),$ $(2,5), (6,5), (7,6), (5,7), (7,8), (6,9), (1,10)\}$. Specifically, $\mathcal{G}(A)$ contains three elementary circuits $\rho = ((2,3), (3,2)), \theta = ((4,3), (3,2), (2,1), (1,4))$, and $\eta = ((6,5), (5,7), (7,6))$. The graph is not strongly connected, for example, $2\mathcal{R}5$, but it does not hold that $5\mathcal{R}2$. In words, node 5 is reachable from node 2, but the converse is not true.

The predecessor and successor sets are given for node 10, for example, by

$$\pi(10) = \{1\}, \ \pi^+(10) = \{1, 2, 3, 4\}, \ \pi^*(10) = \{1, 2, 3, 4, 10\},$$

and

$$\sigma(10) = \sigma^+(10) = \emptyset, \ \sigma^*(10) = \{10\}.$$

For node 5 these sets read

$$\pi(5) = \{2, 6\}, \ \pi^+(5) = \pi^*(5) = \{1, 2, 3, 4, 5, 6, 7\},$$

and

$$\sigma(5) = \{7\}, \ \sigma^+(5) = \sigma^*(5) = \{5, 6, 7, 8, 9\}.$$

There are five m.s.c.s.'s in $\mathcal{G}(A)$ with the set of nodes $[1] = [2] = [3] = [4] = \{1, 2, 3, 4\}$, $[5] = [6] = [7] = \{5, 6, 7\}$, $[8] = \{8\}$, $[9] = \{9\}$, and $[10] = \{10\}$. Because $|\rho|_1 = 2$ and $|\theta|_1 = 4$, the m.s.c.s. corresponding to, for instance, $[2]$ has cyclicity 2, being the greatest common divisor of all circuit lengths in $[2]$. Because $|\eta|_1 = 3$, the m.s.c.s. corresponding to, say $[5]$, has cyclicity 3. The other m.s.c.s.'s have cyclicity 1 by definition. Therefore, the graph $\mathcal{G}(A)$ has cyclicity 6, being the least common multiple of the cyclicity of all m.s.c.s.'s. Hence, $\sigma_{\mathcal{G}(A)} = 6$.
The reduced graph is depicted in Figure 2.3.

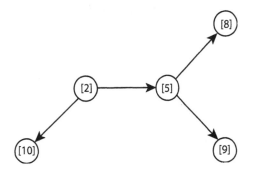

Figure 2.3: The reduced graph of $\mathcal{G}(A)$ in Example 2.1.3.

Based on the reduced graph, let $[i_1] = [8] = \{8\}$, $[i_2] = [9] = \{9\}$, $[i_3] = [5] = \{5, 6, 7\}$, $[i_4] = [10] = \{10\}$, and $[i_5] = [2] = \{1, 2, 3, 4\}$. The corresponding matrices are

$$A_{11} = A_{22} = A_{44} = \varepsilon, \ A_{33} = \begin{pmatrix} \varepsilon & -5 & \varepsilon \\ \varepsilon & \varepsilon & 0 \\ 9 & \varepsilon & \varepsilon \end{pmatrix}, \ A_{55} = \begin{pmatrix} \varepsilon & 0 & \varepsilon & \varepsilon \\ \varepsilon & \varepsilon & -3 & \varepsilon \\ \varepsilon & 4 & \varepsilon & 0 \\ 0 & \varepsilon & \varepsilon & \varepsilon \end{pmatrix}.$$

If both the rows and columns of A are placed into the order

$$8, 9, 5, 6, 7, 10, 1, 2, 3, 4,$$

obtained from placing the elements of the sets $[i_1]$, $[i_2]$, $[i_3]$, $[i_4]$, and $[i_5]$ one after *another, then the following normal form of A is the result*

$$
\begin{pmatrix}
\varepsilon & \varepsilon & \varepsilon & \varepsilon & \frac{1}{2} & \varepsilon & \varepsilon & \varepsilon & \varepsilon & \varepsilon \\
\varepsilon & \varepsilon & \varepsilon & 6 & \varepsilon & \varepsilon & \varepsilon & \varepsilon & \varepsilon & \varepsilon \\
\varepsilon & \varepsilon & \varepsilon & -5 & \varepsilon & \varepsilon & \varepsilon & 16 & \varepsilon & \varepsilon \\
\varepsilon & \varepsilon & \varepsilon & \varepsilon & 0 & \varepsilon & \varepsilon & \varepsilon & \varepsilon & \varepsilon \\
\varepsilon & \varepsilon & 9 & \varepsilon & \varepsilon & \varepsilon & \varepsilon & \varepsilon & \varepsilon & \varepsilon \\
\varepsilon & \varepsilon & \varepsilon & \varepsilon & \varepsilon & \varepsilon & 9 & \varepsilon & \varepsilon & \varepsilon \\
\varepsilon & \varepsilon & \varepsilon & \varepsilon & \varepsilon & \varepsilon & \varepsilon & 0 & \varepsilon & \varepsilon \\
\varepsilon & \varepsilon & \varepsilon & \varepsilon & \varepsilon & \varepsilon & \varepsilon & \varepsilon & -3 & \varepsilon \\
\varepsilon & \varepsilon & \varepsilon & \varepsilon & \varepsilon & \varepsilon & \varepsilon & 4 & \varepsilon & 0 \\
\varepsilon & \varepsilon & \varepsilon & \varepsilon & \varepsilon & \varepsilon & 0 & \varepsilon & \varepsilon & \varepsilon
\end{pmatrix}.
$$

In particular, the diagonal blocks of this normal form of A, starting in the upper left corner and going down to the lower right corner, are given by A_{11}, A_{22}, A_{33}, A_{44}, and A_{55}, respectively.

2.2 EIGENVALUES AND EIGENVECTORS

We start with the definition of one of the most important notions in this book.

DEFINITION 2.4 *Let $A \in \mathbb{R}_{\max}^{n \times n}$ be a square matrix. If $\mu \in \mathbb{R}_{\max}$ is a scalar and $v \in \mathbb{R}_{\max}^{n}$ is a vector that contains at least one finite element such that*

$$
A \otimes v = \mu \otimes v,
$$

then μ is called an eigenvalue *of A and v an* eigenvector *of A associated with eigenvalue μ.*

Note that Definition 2.4 allows an eigenvalue to be $\varepsilon = -\infty$. Also, eigenvectors are allowed to have elements equal to ε as long as they contain finite elements. Note further that a square matrix may have more than one eigenvalue; i.e., eigenvalues are not necessarily unique. Eigenvectors are certainly not unique. Indeed, if v is an eigenvector, then so is $\alpha \otimes v$, with α an arbitrary finite number. Furthermore, observe that the set of all eigenvectors associated with an eigenvalue is a vector space in the max-plus sense, called the eigenspace. Indeed, if v and w are eigenvectors of A associated with eigenvalue μ and α, β are numbers in $\mathbb{R}_{\max}$, then

$$
\begin{aligned}
A \otimes (\alpha \otimes v \oplus \beta \otimes w) &= \alpha \otimes A \otimes v \oplus \beta \otimes A \otimes w \\
&= \alpha \otimes \mu \otimes v \oplus \beta \otimes \mu \otimes w \\
&= \mu \otimes (\alpha \otimes v \oplus \beta \otimes w).
\end{aligned}
$$

Then, if at least one of the numbers α and β is finite, the vector $\alpha \otimes v \oplus \beta \otimes w$ is an eigenvector of A for the eigenvalue μ. The eigenspace of matrix A associated to an eigenvalue μ is denoted by $V(A, \mu)$. When it is clear from the context that the eigenvalue of matrix A is unique, then this eigenvalue often will be denoted by $\lambda(A)$ and the corresponding eigenspace will in that case be denoted by $V(A)$.

A first observation on eigenvalues and eigenvectors is that any finite eigenvalue μ of a square matrix A is the average weight of some circuit in $\mathcal{G}(A)$. To see this, notice that by definition an associated eigenvector v has at least one finite element; that is, there exists a node/index $\eta_1 \in \underline{n}$ such that $v_{\eta_1} \neq \varepsilon$. Then $[A \otimes v]_{\eta_1} = \mu \otimes v_{\eta_1} \neq \varepsilon$. Hence, there exists a node η_2 with

$$a_{\eta_1\eta_2} \otimes v_{\eta_2} = \mu \otimes v_{\eta_1},$$

implying that $a_{\eta_1\eta_2} \neq \varepsilon$, $v_{\eta_2} \neq \varepsilon$, and $(\eta_2, \eta_1) \in \mathcal{D}(A)$. Following the same reasoning, a node η_3 can be found such that

$$a_{\eta_2\eta_3} \otimes v_{\eta_3} = \mu \otimes v_{\eta_2},$$

with $a_{\eta_2\eta_3} \neq \varepsilon$, $v_{\eta_3} \neq \varepsilon$, and $(\eta_3, \eta_2) \in \mathcal{D}(A)$. Proceeding in this way, eventually some node, say, node η_h, must be encountered for a second time, because the number of nodes is finite. We have then found a circuit

$$\gamma = ((\eta_h, \eta_{h+l-1}), (\eta_{h+l-1}, \eta_{h+l-2}), \ldots, (\eta_{h+1}, \eta_h))$$

of length

$$|\gamma|_l = l, \tag{2.3}$$

with weight

$$|\gamma|_w = \bigotimes_{k=0}^{l-1} a_{\eta_{h+k}\eta_{h+k+1}}, \tag{2.4}$$

where $\eta_h = \eta_{h+l}$. By construction,

$$\bigotimes_{k=0}^{l-1} \left(a_{\eta_{h+k}\eta_{h+k+1}} \otimes v_{\eta_{h+k+1}} \right) = \mu^{\otimes l} \otimes \bigotimes_{k=0}^{l-1} v_{\eta_{h+k}}.$$

Recall that $\otimes$ reads as $+$ in conventional algebra. Hence, the above equation reads

$$\sum_{k=0}^{l-1} \left(a_{\eta_{h+k}\eta_{h+k+1}} + v_{\eta_{h+k+1}} \right) = l \times \mu + \sum_{k=0}^{l-1} v_{\eta_{h+k}}. \tag{2.5}$$

Because $\eta_h = \eta_{h+l}$ it follows that

$$\sum_{k=0}^{l-1} v_{\eta_{h+k+1}} = \sum_{k=0}^{l-1} v_{\eta_{h+k}}.$$

Subtracting $\sum_{k=0}^{l-1} v_{\eta_{h+k}}$ on both sides of equation (2.5) yields

$$\sum_{k=0}^{l-1} a_{\eta_{h+k}\eta_{h+k+1}} = l \times \mu,$$

which by (2.4) means that

$$|\gamma|_w = l \times \mu = \mu^{\otimes l}.$$

By (2.3), the average weight of the circuit γ then equals

$$\frac{|\gamma|_w}{|\gamma|_l} = \frac{1}{l} \times \mu^{\otimes l} = \mu.$$

We have thus proved the following lemma.

LEMMA 2.5 *Let $A \in \mathbb{R}_{\max}^{n \times n}$ have a finite eigenvalue μ. Then, a circuit γ exists in $\mathcal{G}(A)$ such that*

$$\mu = \frac{|\gamma|_w}{|\gamma|_1}.$$

According to the above lemma, average weights of circuits are candidates for eigenvalues. Unfortunately, the above lemma does not tell which circuits actually define an eigenvalue. So, why not try the maximal average circuit weight as a first candidate for an eigenvalue? This choice has the advantage that it is independent of any a priori knowledge about particular circuits. In the following this idea is pursued further.

Let $\mathcal{C}(A)$ denote the set of all elementary circuits in $\mathcal{G}(A)$ and write

$$\lambda = \max_{p \in \mathcal{C}(A)} \frac{|p|_w}{|p|_1} \tag{2.6}$$

for the maximal average circuit weight. Notice that $\mathcal{C}(A)$ is a finite set, and if not empty, the maximum on the right-hand side in (2.6) is thus attained by (at least) one circuit in $\mathcal{G}(A)$. In the case where $\mathcal{C}(A) = \emptyset$, define $\lambda = -\infty$. Notice that if A is irreducible, then λ is finite (irreducibility of A implies that $\mathcal{G}(A)$ contains at least one circuit). Note that if $\mathcal{G}(A)$ contains no circuit (for example, if A is a strictly lower triangular matrix), then, according to this definition, $\lambda = -\infty$.

A circuit p in $\mathcal{G}(A)$ is called *critical* if its average weight is maximal, that is, if $\lambda = |p|_w/|p|_1$. The *critical graph* of A, denoted by $\mathcal{G}^c(A) = (\mathcal{N}^c(A), \mathcal{D}^c(A))$, is the graph consisting of those nodes and arcs that belong to critical circuits in $\mathcal{G}(A)$. A node $i \in \mathcal{N}^c(A)$ will sometimes be referred to as a *critical node*. Similarly, a subpath of a critical circuit will be occasionally called a *critical path*. Note that the critical graph of an irreducible matrix does not have to be strongly connected.

Example 2.2.1 *Revisit the situation put forward in Example 2.1.3. The graph $\mathcal{G}(A)$ contains three circuits with average weight $1/2$, $-3/4$, and $4/3$, respectively. The maximal average circuit weight is therefore equal to $4/3$, and the critical graph consists of the circuit $\eta = ((6, 5), (5, 7), (7, 6))$.*

Let A be a square matrix over $\mathbb{R}_{\max}$ whose communication graph contains at least one circuit. Let A' be the matrix obtained from A by subtracting $\tau \in \mathbb{R}$ from every finite element of A. Then the communication graphs of A and A' are the same, except for the arc weights. Furthermore, if a circuit in $\mathcal{G}(A)$ has average weight α, then the same circuit in $\mathcal{G}(A')$ has average weight $\alpha - \tau$. It follows that a circuit in $\mathcal{G}(A)$ is critical if and only if it is critical in $\mathcal{G}(A')$. Hence, the critical graphs $\mathcal{G}^c(A)$ and $\mathcal{G}^c(A')$ are the same, again except for the arc weights. Clearly, by taking τ equal to the maximal average circuit weight in $\mathcal{G}(A)$, the maximal average circuit weight in $\mathcal{G}(A')$ becomes zero.

LEMMA 2.6 *Assuming that $\mathcal{G}(A)$ contains at least one circuit, it follows that any circuit in $\mathcal{G}^c(A)$ is critical.*

Proof. The proof will be given by contradiction. Take λ as defined in (2.6). As indicated above, it may, for ease of exposition, be assumed that $\lambda = 0$. Suppose that $\mathcal{G}^c(A)$ contains a circuit ρ with average weight different from zero. Note that ρ is also a circuit in $\mathcal{G}(A)$. If the average weight of ρ is larger than zero, then the maximal average circuit weight of A is larger than zero, which contradicts the starting point that $\lambda = 0$. Now suppose that the average weight of ρ is negative. Observe that ρ is the concatenation of paths ρ_i, i.e., $\rho = \rho_1 \circ \rho_2 \circ \cdots \circ \rho_\kappa$, where each ρ_i is a subpath of a critical circuit c_i, $i \in \underline{\kappa}$. Hence, there exist subpaths ξ_i such that $c_i = \xi_i \circ \rho_i$, $i \in \underline{\kappa}$. Since all circuits c_i, $i \in \underline{\kappa}$, have weight zero, the circuit composed of the concatenation of the ξ_i's (i.e., $\xi = \xi_1 \circ \xi_2 \circ \cdots \circ \xi_\kappa$) is thus a circuit with positive (average) weight, which contradicts again the starting point that $\lambda = 0$. □

If not stated otherwise, we adopt throughout the text the convention that for any $x \in \mathbb{R}$

$$\varepsilon = \varepsilon - x \qquad \text{and} \qquad \varepsilon - \varepsilon = e. \tag{2.7}$$

Let λ, defined in (2.6), be finite, and consider the matrix A_λ with elements

$$[A_\lambda]_{ij} = a_{ij} - \lambda. \tag{2.8}$$

Matrix A_λ is occasionally referred to as the *normalized* matrix. It is clear that the maximum average circuit weight of $\mathcal{G}(A_\lambda)$ is zero. Therefore, Lemma 2.2 implies that A_λ^+ is well defined, where A_λ^+ should be read as $(A_\lambda)^+$. As noticed before, the set of critical circuits of A and A_λ coincide, and consequently, $\mathcal{G}^c(A)$ and $\mathcal{G}^c(A_\lambda)$ coincide except for their weights. This gives

$$\forall \eta \in \mathcal{N}^c(A) : \quad [A_\lambda^+]_{\eta\eta} = e = 0. \tag{2.9}$$

Indeed, every node of the critical graph is contained in a circuit and every circuit of the critical graph has weight zero. So, any path from a node in the critical graph to itself has weight zero. Next, define

$$A_\lambda^* \stackrel{\text{def}}{=} E \oplus A_\lambda^+ = \bigoplus_{k \geq 0} A_\lambda^{\otimes k}, \tag{2.10}$$

where A_λ^* stands for $(A_\lambda)^*$, and notice that

$$A_\lambda^+ = A_\lambda \otimes (E \oplus A_\lambda^+) = A_\lambda \otimes A_\lambda^*. \tag{2.11}$$

Let $[B]._k$ denote the kth column of a matrix B. The definition of A_λ^* implies that

$$[A_\lambda^*]._\eta = [E \oplus A_\lambda^+]._\eta . \tag{2.12}$$

It follows from (2.12) that the ith element of the vector $[A_\lambda^*]._\eta$ satisfies

$$[A_\lambda^*]_{i\eta} = [E \oplus A_\lambda^+]_{i\eta} = \begin{cases} \varepsilon \oplus [A_\lambda^+]_{i\eta} & \text{for } i \neq \eta, \\ e \oplus [A_\lambda^+]_{i\eta} & \text{for } i = \eta. \end{cases}$$

Then from (2.9) for $\eta \in \mathcal{N}^c(A)$ it follows that

$$[A_\lambda^+]._\eta = [A_\lambda^*]._\eta.$$

If we replace A_λ^+ by $A_\lambda \otimes A_\lambda^*$ (see (2.11)), then the above equality is equivalent to

$$[A_\lambda \otimes A_\lambda^*]._\eta = [A_\lambda^*]._\eta,$$

which gives

$$A_\lambda \otimes [A_\lambda^*]_{\cdot\eta} = [A_\lambda^*]_{\cdot\eta}$$

or, equivalently,

$$A \otimes [A_\lambda^*]_{\cdot\eta} = \lambda \otimes [A_\lambda^*]_{\cdot\eta}.$$

Hence, it follows that λ is an eigenvalue of A and that the ηth column of A_λ^* is an associated eigenvector, for any $\eta \in \mathcal{N}^c(A)$. We summarize our analysis as follows.

LEMMA 2.7 *Let the communication graph $\mathcal{G}(A)$ of matrix $A \in \mathbb{R}_{\max}^{n \times n}$ have finite maximal average circuit weight λ. Then, the scalar λ is an eigenvalue of A, and the column $[A_\lambda^*]_{\cdot\eta}$ is an eigenvector of A associated with λ, for any node η in $\mathcal{G}^c(A)$.*

Lemma 2.7 establishes the existence of an eigenvalue and of associated eigenvectors provided that the maximal average circuit weight is indeed finite. Notice that the irreducibility of A already implies that the maximal average circuit weight is finite. As we will show next, the irreducibility of A moreover implies that any eigenvector associated with any finite eigenvalue of A has only finite elements.

Let v be an eigenvector of A associated with eigenvalue μ, and call the set of nodes of $\mathcal{G}(A)$ corresponding to finite entries of v the *support* of v. Suppose that the support of v does not cover the whole node set of $\mathcal{G}(A)$. If A is irreducible, then any node can be reached from any node and there have to be arcs from the nodes in the support of v going to nodes not belonging to the support of v. Hence, there exists a node j in the support of v and a node i not in the support of v with $a_{ij} \neq \varepsilon$. Then, $a_{ij} \neq \varepsilon$ implies that $[A \otimes v]_i \geq a_{ij} \otimes v_j > \varepsilon$, and the support of $A \otimes v$ is thus bigger than the support of v. Since $\mu \otimes v = A \otimes v$, this contradicts the fact that the support of v and $\mu \otimes v$ have to be equal for any finite μ. We summarize the above analysis as follows.

LEMMA 2.8 *Let $A \in \mathbb{R}_{\max}^{n \times n}$. If A is irreducible, then any vector $v \in \mathbb{R}_{\max}^n$, with at least one finite element, that solves*

$$\mu \otimes v = A \otimes v$$

for some finite μ has all elements different from ε.

Example 2.2.2 *Consider*

$$A = \begin{pmatrix} e & \varepsilon \\ 1 & e \end{pmatrix}.$$

The powers of A are as follows

$$A^{\otimes 2} = \begin{pmatrix} e & \varepsilon \\ 1 & e \end{pmatrix}, \qquad A^{\otimes 3} = \begin{pmatrix} e & \varepsilon \\ 1 & e \end{pmatrix}, \dots,$$

and we conclude that

$$A^{\otimes(k+1)} = e \otimes A^{\otimes k}, \qquad k \geq 1.$$

Hence, we obtain $\lambda = e \, (= 0)$ *as an eigenvalue of A. Moreover, it follows that*

$$A_\lambda \; = \; A \; = \; A_\lambda^* \; = \; A_\lambda^+.$$

Hence, in accordance with Lemma 2.7,

$$\begin{pmatrix} e \\ 1 \end{pmatrix} \qquad and \qquad \begin{pmatrix} \varepsilon \\ e \end{pmatrix}$$

are eigenvectors of A associated with λ. *Notice that A is not irreducible and Lemma 2.8 does not apply, as is illustrated by the eigenvector* $(\varepsilon, e)^\top$, *which has a nonfinite element.*

The existence of a finite eigenvalue has been shown in Lemma 2.7 for the case where A is irreducible. The next step is to show that irreducibility also implies that the eigenvalue of A is unique. Pick any circuit $\gamma = ((\eta_1, \eta_2), (\eta_2, \eta_3), \ldots, (\eta_l, \eta_{l+1}))$ in $\mathcal{G}(A)$ of length $l = |\gamma|_l$ with $\eta_{l+1} = \eta_1$. Then,

$$a_{\eta_{k+1}\eta_k} \neq \varepsilon, \qquad k \in \underline{l}. \tag{2.13}$$

Suppose that μ is a finite eigenvalue of A, and let v be an eigenvector associated with μ. Because it is assumed that $\mu \otimes v = A \otimes v$, it follows that

$$a_{\eta_{k+1}\eta_k} \otimes v_{\eta_k} \leq \mu \otimes v_{\eta_{k+1}}, \qquad k \in \underline{l}.$$

Now argue as in the proof of Lemma 2.5, except that the equalities are replaced by appropriate inequalities. See the text before Lemma 2.5. Proceeding in the above way, we find that the average weight of the circuit γ satisfies

$$\frac{|\gamma|_w}{|\gamma|_l} \; \leq \; \frac{1}{l} \times \mu^{\otimes l} \; = \; \mu.$$

The above analysis holds for any circuit $\gamma \in \mathcal{C}(A)$. In other words, any finite eigenvalue μ has to be larger than or equal to the maximal average circuit weight λ. But, by Lemma 2.5, any finite eigenvalue μ can always be obtained as the average weight of a circuit in $\mathcal{G}(A)$. Hence, λ is a finite eigenvalue of A, and by (2.6) it is uniquely determined.

Suppose now that ε is an eigenvalue of A with corresponding eigenvector v. Then, v has at least one finite element, say, v_η. If A is irreducible, then there is a row γ of A such that $a_{\gamma\eta}$ is finite, which gives

$$\varepsilon = [\varepsilon \otimes v]_\gamma = [A \otimes v]_\gamma \geq a_{\gamma\eta} \otimes v_\eta.$$

But the expression on the above right-hand side is finite, and we conclude that ε cannot be an eigenvalue of an irreducible matrix. Consequently, λ is the unique eigenvalue of A.

For easy reference we summarize our analysis in the following theorem.

THEOREM 2.9 *Any irreducible matrix* $A \in \mathbb{R}_{\max}^{n \times n}$ *possesses one and only one eigenvalue. This eigenvalue, denoted by* $\lambda(A)$, *is a finite number and equal to the maximal average weight of circuits in* $\mathcal{G}(A)$, *i.e.,*

$$\lambda(A) \; = \; \max_{\gamma \in \mathcal{C}(A)} \frac{|\gamma|_w}{|\gamma|_l}.$$

Example 2.2.3   *Revisit the situation put forward in Example 2.1.1. The communication graph of A, given in Figure 2.1, is strongly connected, and A is therefore irreducible. Elaborating on the notation already introduced in Example 2.1.1,
we obtain $C(A) = \{\rho, \theta\}$. The average circuit weights are $|\rho|_w/|\rho|_1 = 13$ and
$|\theta|_w/|\theta|_1 = 12$. Theorem 2.9 applies, and we obtain $\lambda(A) = \max(13, 12) = 13$
for the eigenvalue of A.*

Theorem 2.9 characterizes the eigenvalue of an irreducible square matrix. Algorithms for computing the eigenvalue will be presented in Chapters 5 and 6.

2.3 SOLVING LINEAR EQUATIONS

Following (2.2) and (2.10), we formally define for any $A \in \mathbb{R}_{\max}^{n \times n}$

$$A^* \stackrel{\text{def}}{=} E \oplus A^+ = \bigoplus_{k \geq 0} A^{\otimes k}. \tag{2.14}$$

From Lemma 2.2 it follows easily that A^* exists for any square matrix A with a
communication graph $\mathcal{G}(A)$ having only nonpositive circuit weights. Note that $A^{\otimes n}$
refers to the maximal weight of paths of length n. Hence, these paths contain at least
one circuit. If all circuits have nonpositive circuit weight, then

$$[A^{\otimes n}]_{ij} \leq \bigotimes_{k=0}^{n-1} [A^{\otimes k}]_{ij}, \qquad i, j \in \underline{n},$$

and under the conditions put forward in Lemma 2.2, A^* can be determined as

$$A^* = \bigoplus_{k=0}^{n-1} A^{\otimes k}. \tag{2.15}$$

As will be shown in the next theorem, the operator $(\cdot)^*$ provides the means for
solving the equation $x = A \otimes x \oplus b$, with $A \in \mathbb{R}_{\max}^{n \times n}$ and $b \in \mathbb{R}_{\max}^n$. The precise
statement is as follows.

THEOREM 2.10 *Let $A \in \mathbb{R}_{\max}^{n \times n}$ and $b \in \mathbb{R}_{\max}^n$. If the communication graph $\mathcal{G}(A)$
has maximal average circuit weight less than or equal to e, then the vector $x =
A^* \otimes b$ solves the equation $x = (A \otimes x) \oplus b$. Moreover, if the circuit weights in
$\mathcal{G}(A)$ are negative, then the solution is unique.*

Proof. It will be shown that

$$A^* \otimes b = A \otimes (A^* \otimes b) \oplus b.$$

By Lemma 2.2, A^* exists, implying that

$$A^* \otimes b = \bigoplus_{k \geq 0} A^{\otimes k} \otimes b$$

$$= \left(\bigoplus_{k \geq 1} A^{\otimes k} \otimes b \right) \oplus (E \otimes b)$$

$$= A \otimes \left(\bigoplus_{k \geq 0} A^{\otimes k} \otimes b \right) \oplus (E \otimes b)$$

$$= A \otimes (A^* \otimes b) \oplus b,$$

which concludes the proof of the first part of the theorem.

In order to prove uniqueness under the condition that circuits have negative average weights, argue as follows. Suppose that x is a solution of $x = b \oplus (A \otimes x)$; then substituting the expression for x in $b \oplus (A \otimes x)$ it follows that

$$x = b \oplus (A \otimes b) \oplus (A^{\otimes 2} \otimes x),$$

and repeating the argument, one obtains

$$x = b \oplus (A \otimes b) \oplus (A^{\otimes 2} \otimes b) \oplus (A^{\otimes 3} \otimes x)$$

$$= b \oplus (A \otimes b) \oplus \cdots \oplus (A^{\otimes(k-1)} \otimes b) \oplus (A^{\otimes k} \otimes x)$$

$$= \bigoplus_{l=0}^{k-1} (A^{\otimes l} \otimes b) \oplus (A^{\otimes k} \otimes x). \tag{2.16}$$

The entries of $A^{\otimes k}$ are the maximal weights of paths of length k. For k large enough, any path necessarily contains one or more copies of certain elementary circuits as subpaths, and as k tends to ∞, the number of required elementary circuits tends to ∞. Since circuits have negative weight, the elements of $A^{\otimes k}$ tend to ε, as k tends to ∞, i.e.,

$$\lim_{k \to \infty} A^{\otimes k} \otimes x = \mathcal{E}.$$

Hence, letting k tend to ∞ in equation (2.16) yields that $x = A^* \otimes b$, where Lemma 2.2 is used to show that

$$\lim_{k \to \infty} \bigoplus_{l=0}^{k-1} (A^{\otimes l} \otimes b) = \left(\lim_{k \to \infty} \bigoplus_{l=0}^{k-1} A^{\otimes l} \right) \otimes b = A^* \otimes b.$$

$\square$

To conclude this section, let us consider a different kind of linear equation, namely, $A \otimes x = b$, where A is not necessarily square! A study shows that a solution of this equation does not always exist. However, it will be shown below that one can always find a *greatest solution* to the max-plus inequality $A \otimes x \leq b$. This greatest solution is called the *principal solution* and will be denoted as $x^*(A, b)$.

THEOREM 2.11 *For $A \in \mathbb{R}_{\max}^{m \times n}$ and $b \in \mathbb{R}^m$ it holds that*

$$[x^*(A, b)]_j = \min\{b_i - a_{ij} : i \in \underline{m}\},$$

for $j \in \underline{n}$.

Proof. Notice that if $A \otimes x \leq b$, then the following equivalences hold:

$$\forall i \, \forall j : \quad a_{ij} + x_j \leq b_i$$
$$\Leftrightarrow \forall i \, \forall j : \quad x_j \leq b_i - a_{ij}$$
$$\Leftrightarrow \forall j : \quad x_j \leq \min\{b_i - a_{ij} : i \in \underline{m}\}.$$

$\square$

Please note that in Theorem 2.11 a requirement is that $b \in \mathbb{R}^m$. This theorem may not hold for $b \in \mathbb{R}^m_{\max}$. A counterexample is the scalar equation $\varepsilon \otimes x = \varepsilon$.

From Theorem 2.11 it follows that $x^*(A, b)$, seen as greatest solution of the inequality $A \otimes x \leq b$, is uniquely determined. Indeed, any other solution x of the inequality is such that $x \leq x^*(A, b)$. This situation is different as far as the inequality $A \otimes x \geq b$ is concerned. For such an inequality a *smallest solution* need not be uniquely determined nor even exist.

By Theorem 2.11, any vector x with $x \leq x^*(A, b)$ satisfies $A \otimes x \leq b$. Notice that if A has a column, say, column j, with all elements equal to ε, then $[x^*(A, b)]_j = +\infty$, which means that $A \otimes x \leq b$ can be achieved by a vector whose jth component may take any value in $\mathbb{R}_{\max}$.

Let A be a matrix with the travel times of a train network. Suppose that the planned departure times are given by b; then $x^*(A, b)$ gives the latest departure times of trains from the previous stations such that b still can be met. This kind of *just-in-time* application will be dealt with in more detail in Section 9.1.

Theorem 2.11 is a special result coming from the so-called residuation theory by which optimal solutions can be obtained for inequalities that are generalizations of the above max-plus inequality $A \otimes x \leq b$. A recent and thorough reference on the subject of residuation theory is [25].

2.4 EXERCISES

1. Show that for any $k \geq 1$
$$\lambda^{\otimes k} \otimes A_\lambda^{\otimes k} = A^{\otimes k}.$$

2. Show that for any $k \geq 1$
$$A_\lambda^{\otimes k} = \left(A^{\otimes k}\right)_\nu,$$
with $\nu = k \times \lambda$ (conventional multiplication).

3. Show that if $A \in \mathbb{R}^{n \times n}_{\max}$ is irreducible, then A is regular.

4. Let $A \in \mathbb{R}^{n \times n}_{\max}$ be irreducible. Show that for any node i in $\mathcal{G}(A)$, it holds that $\pi^*(i) = \pi^+(i) = \sigma^+(i) = \sigma^*(i) = \underline{n}$.

5. Let λ denote the maximal average circuit weight in the communication graph of a square matrix A. Consider A_λ and show that
$$0 = \max_{p \in \mathcal{C}(A_\lambda)} \frac{|p|_w}{|p|_l}.$$

6. Let $A \in \mathbb{R}^{n \times n}_{\max}$ be a strictly lower triangular matrix, and consider A^* defined in equation (2.14). Give a direct proof for the fact that A^* and A are related as in equation (2.15). Show that the above need not to be true if A is just lower triangular.

7. Consider the graph depicted in Figure 2.1. What is the maximum weight of a path from 1 to 3 of length 4?

8. Compute eigenvectors for matrix A given in Example 2.1.1.

9. Let

$$A = \begin{pmatrix} \varepsilon & e \\ \varepsilon & e \end{pmatrix}.$$

Show that A has eigenvalues ε and e, and give one corresponding eigenvector for each eigenvalue.

10. For the next matrices A, investigate the existence and uniqueness of a solution of the equation $x = A \otimes x \oplus b$ with $b = \mathbf{u}$. If a solution exists, give the complete solution set of the equation.

$$A = \begin{pmatrix} -1 & e \\ -2 & -1 \end{pmatrix}, \quad A = \begin{pmatrix} e & -2 \\ 2 & -1 \end{pmatrix}, \quad A = \begin{pmatrix} e & e \\ 2 & -1 \end{pmatrix}$$

11. Assume that

$$A = \begin{pmatrix} 1 & 2 \\ e & 1 \end{pmatrix} \quad \text{and} \quad b = \begin{pmatrix} e \\ 1 \end{pmatrix}.$$

Compute $x^*(A, b)$, as defined in Theorem 2.11, and verify that this vector is the greatest solution of the inequality $A \otimes x \leq b$. Show that for the inequality $A \otimes x \geq b$ no (unique) smallest solution exists.

2.5 NOTES

In conventional algebra, a square matrix is called *irreducible* if no identical permutation of its rows and columns exists such that the matrix is transformed into a block upper triangular structure. As shown in [5], the definition in the current chapter of the irreducibility of a square max-plus matrix is equivalent to a max-plus version of the conventional definition.

Theorem 2.9 is the max-plus analogue of the Perron-Frobenius theorem in conventional linear algebra, which states that an irreducible square nonnegative matrix, say, B has a largest eigenvalue that is positive and real, where *largest* means largest in modulus. It is well known that this eigenvalue is given by $\limsup_{k \to \infty} \left(\text{tr}(B^k) \right)^{1/k}$, where $\text{tr}(B^k)$ denotes the trace of the kth power of the nonnegative matrix B in a conventional sense.

To see the parallel with max plus, notice that the maximal average circuit weight of circuits of length k crossing node i is given by $(1/k) \times [A^{\otimes k}]_{ii}$. Theorem 2.9 then yields

$$\lambda = \bigoplus_{k \geq 1} \left(\frac{1}{k} \times \max_{i \in \underline{n}} [A^{\otimes k}]_{ii} \right) = \bigoplus_{k \geq 1} \left(\frac{1}{k} \times \bigoplus_{i=1}^{n} [A^{\otimes k}]_{ii} \right) = \bigoplus_{k \geq 1} \left(\text{tr}_{\oplus}(A^{\otimes k}) \right)^{\otimes (1/k)},$$

where $\text{tr}_{\oplus}(A^{\otimes k}) \stackrel{\text{def}}{=} \bigoplus_{i=1}^{n} [A^{\otimes k}]_{ii}$ stands for the trace of $A^{\otimes k}$ in max-plus sense. Note that $\bigoplus_{k \geq 1}$ can be seen as $\sup_{k \geq 1}$. Circuits of length larger than n can be built up from circuits of length at most n. Therefore, the maximal average weight of these two types of circuits can be expressed in terms of each other. From this it easily follows that for all $h \geq n$

$$\lambda = \sup_{k \geq 1} \left(\text{tr}_{\oplus}(A^{\otimes k}) \right)^{\otimes (1/k)} = \sup_{k \geq h} \left(\text{tr}_{\oplus}(A^{\otimes k}) \right)^{\otimes (1/k)} = \limsup_{h \to \infty} \left(\text{tr}_{\oplus}(A^{\otimes h}) \right)^{\otimes (1/h)},$$

yielding a similar expression for the max-plus eigenvalue as in the conventional case.

The operator $(\cdot)^*$ defined in (2.14) is in the literature referred to as the *Kleene star*. In the present chapter it is shown that the Kleene star of a square matrix over $\mathbb{R}_{\max}$ exists if any cycle weight in its communication graph is nonpositive. The Kleene star again is encountered in Section 4.2, where it is shown that eigenvectors can be characterized through the Kleene star of the normalized matrix, defined in (2.8). See also Lemma 2.7.

Almost twenty years ago, Professor Cuninghame-Green was invited to Delft University of Technology in order to give some lectures about "his" max-plus algebra. We came to talk about the well-known realization problem in mathematical systems theory. The simplest version is as follows. Given a series of scalars g_i, $i \in \mathbb{N}$, find $A \in \mathbb{R}^{n \times n}$, $B \in \mathbb{R}^{n \times 1}$, and $C \in \mathbb{R}^{1 \times n}$ such that $CA^iB = g_i$ for all $i \in \mathbb{N}$, everything in conventional algebra, and in such a way that n is as small as possible. The solution in conventional systems theory is well known, but what can one say about the same problem statement and set of equations in max-plus algebra? Professor Cuninghame-Green did not see the solution immediately. But when he left, thanking us for the hospitality, he seemed rather confident and said he would send us the solution the following week.... In the meantime, many papers with partial results have been published, but to the authors' current knowledge, the general solution is still unknown; see [76] and [38].

Chapter Three

Periodic Behavior and the Cycle-Time Vector

This chapter deals with sequences $\{x(k) : k \in \mathbb{N}\}$ generated by

$$x(k+1) = A \otimes x(k),$$

for $k \geq 0$, where $A \in \mathbb{R}_{\max}^{n \times n}$ and $x(0) = x_0 \in \mathbb{R}_{\max}^n$ is the initial condition. The sequences are then equivalently described by

$$x(k) = A^{\otimes k} \otimes x_0, \tag{3.1}$$

for all $k \geq 0$.

DEFINITION 3.1 *Let $\{x(k) : k \in \mathbb{N}\}$ be a sequence in $\mathbb{R}_{\max}^n$, and assume that for all $j \in \underline{n}$ the quantity η_j, defined by*

$$\lim_{k \to \infty} \frac{x_j(k)}{k},$$

exists. The vector $\eta = (\eta_1, \eta_2, \ldots, \eta_n)^\top$ is called the cycle-time vector *of the sequence $x(k)$. If all η_j's have the same value, this value is also called the* asymptotic growth rate *of the sequence $x(k)$.*

Throughout this chapter the sequences $\{x(k) : k \in \mathbb{N}\}$ will be generated by a recurrence relation as given above, with initial condition x_0. It will be shown that once a cycle-time vector exists, its value is independent of x_0 and is basically determined by matrix A involved; see, for instance, (3.1). For this reason, the vector η will occasionally also be referred to as the *cycle-time vector* of the associated matrix. A similar remark holds with respect to the asymptotic growth rate.

Note that a vector of n identical asymptotic growth rates can be seen as the cycle-time vector in case all the limits $\lim_{k \to \infty} x_j(k)/k$, $j \in \underline{n}$, have the same value. Hence, the notions of cycle-time vector and asymptotic growth rate are closely related. For this reason, the cycle-time vector will occasionally also be referred to as the *asymptotic growth rate*, and vice versa. Hence, the notions cycle-time vector and asymptotic growth rate will be used interchangeably. However, from the context it is always clear which of the two notions, defined in Definition 3.1, is actually meant. Sometimes, even both notions apply.

This chapter deals with the *quantitative* asymptotic behavior of $x(k)$. By quantitative behavior, the cycle-time vector as well as the asymptotic growth rate of $x(k)$ is meant; see Section 1.3 for an application in the heap model. If A is irreducible, with unique eigenvalue λ and associated (finite) eigenvector v, then for $x(0) = v$ it follows that

$$x(k) = A^{\otimes k} \otimes x(0)$$
$$= \lambda^{\otimes k} \otimes v$$

for all $k \geq 0$, which gives for any $j \in \underline{n}$ that

$$\lim_{k \to \infty} \frac{x_j(k)}{k} = \lambda \tag{3.2}$$

and the asymptotic growth rate of $x(k)$ coincides with the eigenvalue of A.

The key question about the limiting behavior is what happens if $x(k)$ is initialized with an *arbitrary finite* vector x_0, not necessarily an eigenvector, or what happens if A is reducible (or not irreducible). This chapter is devoted to answering these questions. Specifically, it will be shown that for regular A the cycle-time vector of $x(k)$ exists and is independent of the initial vector x_0. Further, if A is irreducible, then the asymptotic growth rate of any $x_j(k), j \in \underline{n}$, is equal to the eigenvalue of A.

The chapter is organized as follows. In Section 3.1, a key result is presented characterizing powers of an irreducible matrix by means of its eigenvalue and cyclicity. The cycle-time vector of $x(k)$ for regular matrices is studied in Sections 3.2 and 3.3. In Section 3.4, a special class of reducible matrices is studied that enjoy some nice algebraic properties.

3.1 CYCLICITY AND TRANSIENT TIME

According to Theorem 2.9, any irreducible matrix $A \in \mathbb{R}_{\max}^{n \times n}$ possesses a unique eigenvalue. This section establishes an important statement on the asymptotic behavior of the powers of A in terms of the eigenvalue. Before proving the main result, a number of preliminary technical results will be presented. We start with a fundamental theorem; see, for example, [18].

THEOREM 3.2 *Let $\beta_1, \ldots, \beta_q$ be natural numbers such that their greatest common divisor is one; in symbols, $\gcd\{\beta_1, \ldots, \beta_q\} = 1$. Then, there exists a natural number N such that for all $k \geq N$ there are integers $n_1, \ldots, n_q \geq 0$ such that $k = (n_1 \times \beta_1) + \cdots + (n_q \times \beta_q)$.*

The next result is an important one from graph theory. Below, the communication graph $\mathcal{G}(A)$ of an irreducible matrix $A \in \mathbb{R}_{\max}^{n \times n}$ will be considered. The cyclicity of a graph has been introduced in Chapter 2, and for the graph $\mathcal{G}(A)$ it will be denoted by $\sigma_{\mathcal{G}(A)}$. Recall that $\mathcal{G}(A) = (\mathcal{N}(A), \mathcal{D}(A))$, where $\mathcal{N}(A)$ is the set of nodes and $\mathcal{D}(A)$ is the set of directed edges of $\mathcal{G}(A)$. To simplify the notation, in the following we write $\mathcal{N}$, $\mathcal{D}$, and $\sigma_{\mathcal{G}}$ for the set of nodes, the set of directed arcs, and the cyclicity of $\mathcal{G}(A)$, respectively.

LEMMA 3.3 *Let $A \in \mathbb{R}_{\max}^{n \times n}$ be an irreducible matrix, and let the cyclicity of its communication graph be $\sigma_{\mathcal{G}}$. Then, after a suitable relabeling of the nodes of $\mathcal{G}(A)$, the matrix $A^{\otimes \sigma_{\mathcal{G}}}$ corresponds to a block diagonal matrix with $\sigma_{\mathcal{G}}$ blocks on the diagonal. The communication graph of each diagonal block is strongly connected and has cyclicity one. Moreover, the eigenvalues of all diagonal blocks have the same value.*

Proof. Write $\mathcal{G}(A) = (\mathcal{N}, \mathcal{D})$, and consider the relation between nodes $i, j \in \mathcal{N}$ characterized by

$$i \mathcal{K} j \iff \text{the length of every path from } i \text{ to } j \text{ is a multiple of } \sigma_{\mathcal{G}}. \quad (3.3)$$

It can easily be shown that this relation is an equivalence relation on $\mathcal{N}$. Further, let $k_0 \in \mathcal{N}$ be an arbitrarily chosen, but fixed node; then equivalence classes $\mathcal{C}_0, \mathcal{C}_1, \ldots, \mathcal{C}_{\sigma_{\mathcal{G}}-1}$ associated with the equivalence relation (3.3) can be introduced as

$$i \in \mathcal{C}_l \iff \text{every path from } k_0 \text{ to } i \text{ has length (mod } \sigma_{\mathcal{G}}) \text{ equal to } l, \quad (3.4)$$

for $l = 0, 1, \ldots, \sigma_{\mathcal{G}} - 1$. It is not difficult to show for any $i, j \in \mathcal{N}$ that $i \mathcal{K} j \iff i, j \in \mathcal{C}_l$ for some $l = 0, 1, \ldots, \sigma_{\mathcal{G}} - 1$.

Assume that there is a path from i to j of length $\sigma_{\mathcal{G}}$. Then it follows that every path from i to j has a length that is a multiple of $\sigma_{\mathcal{G}}$. Indeed, concatenation of the previously mentioned paths with one and the same path from j to i yields circuits whose lengths must be multiples of $\sigma_{\mathcal{G}}$. Hence, every path of length $\sigma_{\mathcal{G}}$ must end in the same class as the class from which it starts. Because $A^{\otimes \sigma_{\mathcal{G}}}$ can be computed by considering all paths of length $\sigma_{\mathcal{G}}$, it follows that $A^{\otimes \sigma_{\mathcal{G}}}$ is block diagonal, possibly after an appropriate relabeling of the nodes according to the classes $\mathcal{C}_0, \mathcal{C}_1, \ldots, \mathcal{C}_{\sigma_{\mathcal{G}}-1}$; for instance, by first labeling all nodes in $\mathcal{C}_0$, then all nodes in $\mathcal{C}_1$, and so on.

Further, since for all $i, j \in \mathcal{C}_l$ there is a path from i to j whose length is a multiple of $\sigma_{\mathcal{G}}$, it follows that the block in $A^{\otimes \sigma_{\mathcal{G}}}$ corresponding to class $\mathcal{C}_l$ is irreducible. Indeed, the previous path from i to j can be seen as a concatenation of a number of subpaths, all of length $\sigma_{\mathcal{G}}$ and each going from one node in $\mathcal{C}_l$ to another node in $\mathcal{C}_l$. Now considering all such subpaths of maximal weight, it follows that the communication graph of the block in $A^{\otimes \sigma_{\mathcal{G}}}$ corresponding to class $\mathcal{C}_l$ is strongly connected and that the block itself is irreducible.

Finally, every circuit in $\mathcal{G}(A)$ must go through the equivalence classes $\mathcal{C}_0, \mathcal{C}_1, \ldots, \mathcal{C}_{\sigma_{\mathcal{G}}-1}$. Indeed, suppose there is a circuit going through just τ of the classes, where $\tau < \sigma_{\mathcal{G}}$. Then there must be a class $\mathcal{C}_l$ and nodes $i, j \in \mathcal{C}_l$ such that there is a path from i to j of length less than or equal to τ. However, this is in contradiction with the fact that any path between nodes of the same class must be a multiple of $\sigma_{\mathcal{G}}$. Hence, it follows that the number of circuits in $\mathcal{G}(A)$ is the same as the number of circuits going though any class $\mathcal{C}_l$. Observe that circuits in $\mathcal{G}(A)$ of length $\kappa \times \sigma_{\mathcal{G}}$ can be associated with circuits in $\mathcal{G}(A^{\otimes \sigma_{\mathcal{G}}})$ of length κ. Since the greatest common divisor of all circuit lengths in $\mathcal{G}(A)$ is $\sigma_{\mathcal{G}}$, it follows that the communication graph of the block in $A^{\otimes \sigma_{\mathcal{G}}}$ corresponding to class $\mathcal{C}_l$ has cyclicity one.

The fact that the eigenvalues of the diagonal blocks are identical follows immediately from the irreducibility of A. $\qquad \square$

Example 3.1.1   *Consider the following irreducible matrix and its square,*

$$
A = \begin{pmatrix}
\varepsilon & 0 & \varepsilon & \varepsilon & \varepsilon \\
0 & \varepsilon & \varepsilon & \varepsilon & 1 \\
\varepsilon & 1 & \varepsilon & \varepsilon & \varepsilon \\
\varepsilon & \varepsilon & 1 & \varepsilon & \varepsilon \\
\varepsilon & \varepsilon & \varepsilon & 1 & \varepsilon
\end{pmatrix}, \quad
A^{\otimes 2} = \begin{pmatrix}
0 & \varepsilon & \varepsilon & \varepsilon & 1 \\
\varepsilon & 0 & \varepsilon & 2 & \varepsilon \\
1 & \varepsilon & \varepsilon & \varepsilon & 2 \\
\varepsilon & 2 & \varepsilon & \varepsilon & \varepsilon \\
\varepsilon & \varepsilon & 2 & \varepsilon & \varepsilon
\end{pmatrix}.
$$

It follows that $\sigma_{\mathcal{G}(A)} = 2$. See Figure 3.1. If we take $k_0 = 1$, then it follows from (3.4) that $C_0 = \{1, 3, 5\}$ and $C_1 = \{2, 4\}$. A permutation of the rows and columns of $A^{\otimes 2}$ according to these two classes leads to a block diagonal matrix, being a normal form of $A^{\otimes 2}$,

$$
\left(\begin{array}{ccc|cc}
0 & \varepsilon & 1 & \varepsilon & \varepsilon \\
1 & \varepsilon & 2 & \varepsilon & \varepsilon \\
\varepsilon & 2 & \varepsilon & \varepsilon & \varepsilon \\
\hline
\varepsilon & \varepsilon & \varepsilon & 0 & 2 \\
\varepsilon & \varepsilon & \varepsilon & 2 & \varepsilon
\end{array}\right).
$$

Also, it is easily seen by the finite diagonal elements in both blocks that the graphs of both blocks have cyclicity one.

COROLLARY 3.4 *Under the conditions of Lemma 3.3, let τ be a multiple of $\sigma_{\mathcal{G}(A)}$. Then, after a relabeling of the nodes of $\mathcal{G}(A)$, the matrix $A^{\otimes \tau}$ corresponds to a block diagonal matrix with $\sigma_{\mathcal{G}(A)}$ blocks on the diagonal. The communication graph of each diagonal block is strongly connected and has cyclicity one.*

Proof. The proof may be obtained along the lines of thought of the proof of Lemma 3.3 as far as the block diagonal structure is concerned. For the strong connectedness and cyclicity of the graph of each block, Theorem 3.2 can play a useful role, similar to its application in the proof of Lemma 3.6.   □

DEFINITION 3.5 *Let $A \in \mathbb{R}_{\max}^{n \times n}$ be such that its communication graph contains at least one circuit. The cyclicity of A, denoted by $\sigma(A)$, is the cyclicity of the critical graph of A.*

It may be surprising that the cyclicity of matrix A is defined via its critical graph and not via its communication graph. The reason for this will become clear later when we show that the cyclicity of the critical graph determines the periodic behavior of powers of A.

Suppose that the critical graph of an irreducible matrix A has q maximal strongly connected subgraphs (m.s.c.s.), and let m.s.c.s. i have cyclicity σ_i. Then each σ_i, with $i \in \underline{q}$, is a multiple of $\sigma_{\mathcal{G}(A)}$, and consequently, $\sigma(A)$ is a multiple of $\sigma_{\mathcal{G}(A)}$.

Example 3.1.2   *(Continuation of Example 3.1.1.) It follows that $\sigma(A) = 4$. Also it is clear that the critical graph $\mathcal{G}^c(A)$ does not cover all nodes of the strongly connected graph $\mathcal{G}(A)$. See also Figure 3.1.*

With Definition 3.5 we now have two types of cyclicity in relation to matrix A, namely, the cyclicity of the communication graph of A, denoted as $\sigma_{\mathcal{G}(A)}$ and

Figure 3.1: Communication graph of Example 3.1.1 (left) and the corresponding critical graph (right).

occasionally referred to as the graph cyclicity of A, and the cyclicity of matrix A itself, denoted by $\sigma(A)$ and occasionally referred to as the matrix cyclicity of A. From Definition 3.5 it follows that $\sigma(A) = \sigma_{\mathcal{G}^c(A)}$; i.e., the matrix cyclicity of A is equal to the cyclicity of its critical graph.

With the above terminology, Lemma 3.3 now states that if A is an irreducible matrix with graph cyclicity $\sigma_{\mathcal{G}(A)}$, then, after a suitable relabeling of the nodes of $\mathcal{G}(A)$, $A^{\otimes \sigma_{\mathcal{G}(A)}}$ corresponds to a block diagonal matrix with square diagonal blocks that are irreducible and have graph cyclicity one. Clearly, the latter result is completely in terms of graph cyclicity. However, it will turn out that a similar result can be stated completely in terms of matrix cyclicity. To derive this result, we introduce first some preliminary notation and results.

Let A be an irreducible matrix, and let $\mathcal{G}^c(A)$ be its critical graph. Now let A^c be the submatrix of A such that the communication graph of A^c is equal to the critical graph of A, i.e., $\mathcal{G}(A^c) = \mathcal{G}^c(A)$. Matrix A^c can be obtained from matrix A by restricting A to those entries that correspond to arcs in $\mathcal{G}^c(A)$. Occasionally, matrix A^c is referred to as a critical matrix. Since all circuits in a critical graph are critical (see Lemma 2.6), it follows that the critical graph of A^c is the communication graph of A^c itself, i.e., $\mathcal{G}^c(A^c) = \mathcal{G}(A^c)$, implying $\sigma_{\mathcal{G}^c(A^c)} = \sigma_{\mathcal{G}(A^c)}$. From this it follows that the matrix cyclicity of A^c (i.e., $\sigma(A^c)$) is equal to the graph cyclicity of A^c (i.e., $\sigma_{\mathcal{G}(A^c)}$). Hence, for the critical matrix A^c both types of cyclicity coincide and are equal to $\sigma(A)$. From the above it follows further that $\mathcal{G}(A^c) = \mathcal{G}^c(A) = \mathcal{G}^c(A^c)$. However, we can prove more.

LEMMA 3.6 *Let A be an irreducible matrix, and let A^c be its corresponding critical matrix. Then, for all $k \geq 1$,*

$$\mathcal{G}((A^c)^{\otimes k}) = \mathcal{G}^c(A^{\otimes k}) = \mathcal{G}^c((A^c)^{\otimes k}).$$

Proof. Note that A^c is a submatrix of A and $(A^c)^{\otimes k}$ is a submatrix of $A^{\otimes k}$. Further, note that $\mathcal{G}^c(\cdot)$ is a subgraph of $\mathcal{G}(\cdot)$, and denote this as $\mathcal{G}^c(\cdot) \subseteq \mathcal{G}(\cdot)$. Then it follows that $\mathcal{G}^c((A^c)^{\otimes k}) \subseteq \mathcal{G}^c((A)^{\otimes k})$ and $\mathcal{G}^c((A^c)^{\otimes k}) \subseteq \mathcal{G}((A^c)^{\otimes k})$.

To prove the converse inclusions, note that any arc in $\mathcal{G}(A^{\otimes k})$ from node j to node i can be interpreted as a path in $\mathcal{G}(A)$ of length k from node j to node i. Then, if a number of arcs in $\mathcal{G}(A^{\otimes k})$ form a circuit, say, of length l, then the associated paths in $\mathcal{G}(A)$ form a circuit of length $k \times l$. Conversely, consider a circuit in $\mathcal{G}(A)$, choose an arbitrary node on the circuit, and traverse the circuit with steps of length k until the chosen node is reached again. If l such steps are needed, then there exists a circuit in $\mathcal{G}(A^{\otimes k})$ of length l. In the same way, critical circuits in $\mathcal{G}(A^{\otimes k})$ of length l correspond to critical circuits in $\mathcal{G}(A)$ of length $k \times l$ and conversely.

If ψ is a critical circuit of length l in $\mathcal{G}(A^{\otimes k})$, then there is a corresponding critical circuit ψ' of length $k \times l$ in $\mathcal{G}(A)$. Because ψ' is critical, it is also a circuit in $\mathcal{G}^c(A)$, in turn implying that ψ is a critical circuit in $\mathcal{G}((A^c)^{\otimes k})$. Hence, it follows that $\mathcal{G}^c((A^c)^{\otimes k}) \supseteq \mathcal{G}^c((A)^{\otimes k})$. The remaining inclusion can be proved in a similar way.   □

We now are going to see the consequences of the above for Lemma 3.3 applied to the critical A^c. Therefore, we first assume that the matrix is irreducible and has cyclicity (both graph and matrix) σ. Then, according to Lemma 3.3, $(A^c)^{\otimes \sigma}$, again after a suitable relabeling of nodes, corresponds to a block diagonal matrix with square diagonal blocks that are irreducible and have graph cyclicity one. However, since $\mathcal{G}^c((A^c)^{\otimes \sigma}) = \mathcal{G}((A^c)^{\otimes \sigma})$ (see Lemma 3.6 with $k = \sigma$), the graph of each of the diagonal blocks of $(A^c)^{\otimes \sigma}$ coincides with its critical graph. Thus, for each diagonal block both cyclicities coincide, and therefore both are one.

If A^c is not irreducible, the above can be done for each of m.s.c.s.'s of $\mathcal{G}^c(A)$ with their individual cyclicities (graph and matrix). The least common multiple of these cyclicities equals the matrix cyclicity of A. If this cyclicity is denoted by σ, it follows with Corollary 3.4 (note that σ is a multiple of $\sigma_{\mathcal{G}(A)}$) that each diagonal block of $(A^c)^{\otimes \sigma}$ corresponds to a block diagonal matrix with square diagonal blocks that are irreducible and have cyclicity (both graph and matrix) one. To make the overall block diagonal matrix of the same size as A it possibly has to be augmented with one square block with entries equal to ε. The latter block arises when $\mathcal{G}^c(A)$ does not cover all nodes.

In all cases it follows that each finite diagonal block of the block diagonal matrix corresponding to $(A^c)^{\otimes \sigma}$ is irreducible and has cyclicity (both graph and matrix) one. Taking the least common multiple of all cyclicities, it follows that the cyclicity of the whole matrix $(A^c)^{\otimes \sigma}$ is one, i.e., $\sigma_{\mathcal{G}^c((A^c)^{\otimes \sigma})} = 1$.

With these preliminary observations, the next lemma can easily be proved.

LEMMA 3.7 *Let $A \in \mathbb{R}_{\max}^{n \times n}$ be an irreducible matrix with cyclicity $\sigma = \sigma(A)$. Then, the cyclicity of matrix $A^{\otimes \sigma}$ is equal to one.*

Proof. According to Lemma 3.6 with $k = \sigma$, it follows that the graphs $\mathcal{G}^c(A^{\otimes \sigma})$ and $\mathcal{G}^c((A^c)^{\otimes \sigma})$ are the same. By the above observations, the latter graph has cyclicity one. Hence, the cyclicity of $A^{\otimes \sigma}$, being the cyclicity of the former graph, is also one.   □

Finally, as a last prerequisite for the main result of this section, we now treat a special case.

LEMMA 3.8 *Let $A \in \mathbb{R}_{\max}^{n \times n}$ be an irreducible matrix with eigenvalue e and cyclicity one. Then there is an N such that $A^{\otimes (k+1)} = A^{\otimes k}$ for all $k \geq N$.*

Proof. Let $\mathcal{G}^c(A)$ be the critical graph with node set $\mathcal{N}^c$. It will subsequently be shown that there exists an N such that, for all $k \geq N$:

1. $[A^{\otimes k}]_{ii} = [A^+]_{ii} = e$ for all $i \in \mathcal{N}^c$,

2. $[A^{\otimes k}]_{ij} = [A^+]_{ij}$ for all $i \in \mathcal{N}^c$ and $j \in \underline{n}$,

3. $[A^{\otimes k}]_{ij} = \bigoplus_{l \in \mathcal{N}^c} [A^+]_{il} \otimes [A^+]_{lj}$ for all $i, j \in \underline{n}$.

From these three statements, it follows that $A^{\otimes(k+1)} = A^{\otimes k}$ for all $k \geq N$. Clearly, only statement 3 is required to complete the proof of the lemma. However, statement 2 will be used in proving statement 3, while statement 1 plays a role in the proof of statement 2. The reason for considering the three statements is that they provide a nice structure for the proof of Lemma 3.8. Therefore, to conclude the proof of the lemma, the above three statements will now subsequently be proved.

Statement 1. Consider $i \in \mathcal{N}^c$. Then there is a strongly connected component of $\mathcal{G}^c(A)$, say $\mathcal{G}_1$ with node set $\mathcal{N}_1$, such that $i \in \mathcal{N}_1$. Note that $\mathcal{G}_1$ is a critical (sub)graph, implying that all its circuits are critical. Since the cyclicity of matrix A is one, it further follows that the cyclicity of graph $\mathcal{G}_1$ is equal to one too. Hence, there exist circuits in $\mathcal{G}_1$, say, $\zeta_1, \ldots, \zeta_q$, whose lengths have a greatest common divisor equal to one; in symbols, $\gcd\{|\zeta_1|_1, \ldots, |\zeta_q|_1\} = 1$. Since $\mathcal{G}_1$ is strongly connected, there exists a circuit α in $\mathcal{G}_1$ such that i is a node in α and $\alpha \cap \zeta_j \neq \emptyset$ for $j \in q$; that is, α passes through i and through all circuits $\zeta_1, \ldots, \zeta_q$. Then, for any $n_1, \ldots, n_q \in \mathbb{N}$, there is a circuit passing through i, built from circuit α, n_1 copies of circuit ζ_1, n_2 copies of circuit ζ_2 and so on, up to n_q copies of circuit ζ_q. By Theorem 3.2, it follows that there is an N such that for each $k \geq N$, there exist integers $n_1, \ldots, n_q \in \mathbb{N}$, such that

$$k = |\alpha|_1 + n_1 \times |\zeta_1|_1 + \cdots + n_q \times |\zeta_q|_1.$$

For these $n_1, \ldots, n_q$, construct a circuit passing through i, built from circuit α, n_1 copies of circuit ζ_1, n_2 copies of circuit ζ_2 and so on, up to n_q copies of circuit ζ_q. It is clear that the circuit is one in $\mathcal{G}_1$. Therefore, it is itself also a critical circuit with weight e. Since the maximal circuit mean in $\mathcal{G}(A)$ is e, it follows that $[A^{\otimes k}]_{ii} = e$ for all $k \geq N$, by the definition of $[A^+]_{ii}$, also implying that $[A^+]_{ii} = e$.

Statement 2. By the definition of $[A^+]_{ij}$ there exists an l such that $[A^{\otimes l}]_{ij} = [A^+]_{ij}$. In fact, since the eigenvalue of A is e, it follows by Lemma 2.2 even that $l \leq n$. Then it follows that

$$[A^{\otimes(k+l)}]_{ij} \geq [A^{\otimes k}]_{ii} \otimes [A^{\otimes l}]_{ij} = [A^{\otimes l}]_{ij} = [A^+]_{ij},$$

for k large enough, $i \in \mathcal{N}^c$, and $j \in \underline{n}$; see statement 1. Clearly,

$$[A^+]_{ij} = \bigoplus_{m \geq 1} [A^{\otimes m}]_{ij} \geq [A^{\otimes(k+l)}]_{ij} \geq [A^+]_{ij}.$$

If we replace $k + l$ by k, it therefore follows that $[A^{\otimes k}]_{ij} = [A^+]_{ij}$ for all $i \in \mathcal{N}^c, j \in \underline{n}$, with k large enough. Dually, it follows, of course, that $[A^{\otimes m}]_{ij} = [A^+]_{ij}$ for all $i \in \underline{n}, j \in \mathcal{N}^c$, and m large enough.

Statement 3. Take k and m large enough such that $[A^{\otimes k}]_{il} = [A^+]_{il}$ and $[A^{\otimes m}]_{lj} = [A^+]_{lj}$ for all $l \in \mathcal{N}^c$; see statement 2. Then

$$[A^{\otimes(k+m)}]_{ij} \geq [A^{\otimes k}]_{il} \otimes [A^{\otimes m}]_{lj} = [A^+]_{il} \otimes [A^+]_{lj},$$

for all $l \in \mathcal{N}^c$. If we substitute k for $k + m$, then it follows for k large enough that

$$[A^{\otimes k}]_{ij} \geq \bigoplus_{l \in \mathcal{N}^c} [A^+]_{il} \otimes [A^+]_{lj}.$$

Now consider a path from j to i not passing through $\mathcal{N}^c$. Such a path consists of an elementary path and a number of circuits all having negative weight (i.e., a weight less than e). Let the average weight of a noncritical circuit be maximally δ. Then the weight of a path from j to i of length $k + 1$ not passing through a node in $\mathcal{N}^c$ can be bounded from above by $[A^+]_{ij} + k \times \delta = [A^+]_{ij} \otimes \delta^{\otimes k}$, where $[A^+]_{ij}$ is a fixed upper bound for the weight of the elementary path and $k \times \delta$ is an upper bound for the total weight of the circuits. Since $\delta < e$ (i.e., $\delta < 0$ in conventional notation), it follows that for k large enough

$$[A^+]_{ij} \otimes \delta^{\otimes k} \leq \bigoplus_{l \in \mathcal{N}^c} [A^+]_{il} \otimes [A^+]_{lj} .$$

Indeed, the right-hand side of the inequality is fixed, while the left-hand side tends to $-\infty$ for k going to $+\infty$. Hence, for k large enough it follows that

$$[A^{\otimes k}]_{ij} = \bigoplus_{l \in \underline{n}} [A^+]_{il} \otimes [A^+]_{lj} = \bigoplus_{l \in \mathcal{N}^c} [A^+]_{il} \otimes [A^+]_{lj} ,$$

for all $i, j \in \underline{n}$. $\square$

We now state the celebrated cyclicity theorem of max-plus algebra.

THEOREM 3.9 *Let $A \in \mathbb{R}_{\max}^{n \times n}$ be an irreducible matrix with eigenvalue λ and cyclicity $\sigma = \sigma(A)$. Then there is an N such that*

$$A^{\otimes(k+\sigma)} = \lambda^{\otimes \sigma} \otimes A^{\otimes k}$$

for all $k \geq N$.

Proof. Consider matrix $B = (A_\lambda)^{\otimes \sigma}$. Recall that σ is the cyclicity of the critical graph of A, which is a multiple of the cyclicity of the communication graph of A itself. Then, by Corollary 3.4, after a suitable relabeling of the nodes of $\mathcal{G}(A)$, matrix B is a block diagonal matrix with square diagonal blocks of which the communication graphs are strongly connected and have cyclicity one. By Lemma 3.7 it follows that the matrix cyclicity of B is one, implying that the matrix cyclicity of each of its diagonal blocks is one. Hence, by applying Lemma 3.8 to each diagonal block, it ultimately follows that an M exists such that $B^{\otimes(l+1)} = B^{\otimes l}$, for all $l \geq M$. The latter implies that

$$\left((A_\lambda)^{\otimes \sigma} \right)^{\otimes(l+1)} = \left((A_\lambda)^{\otimes \sigma} \right)^{\otimes l} ,$$

which can be further written as $(A_\lambda)^{\otimes(l \times \sigma + \sigma)} = (A_\lambda)^{\otimes(l \times \sigma)}$ or

$$A^{\otimes(l \times \sigma + \sigma)} = \lambda^{\otimes \sigma} \otimes A^{\otimes(l \times \sigma)},$$

for all $l \geq M$. Finally, note that $A^{\otimes(l \times \sigma + j + \sigma)} = \lambda^{\otimes \sigma} \otimes A^{\otimes(l \times \sigma + j)}$, for any $j, 0 \leq j \leq \sigma - 1$, implying that for all $k \geq N \overset{\text{def}}{=} M \times \sigma$ it follows that

$$A^{\otimes(k+\sigma)} = \lambda^{\otimes \sigma} \otimes A^{\otimes k}.$$

$\square$

Consider matrix A as in Theorem 3.9 with eigenvalue λ and cyclicity $\sigma = \sigma(A)$. Note that then also $\sigma = \sigma(A_\lambda)$. The theorem implies that there is a periodic behavior in the sequence of the powers of A_λ with a length σ equal to the cyclicity of A_λ, i.e., $A_\lambda^{\otimes(k+\sigma)} = A_\lambda^{\otimes k}$ for k large enough. With the ideas put forward in Section 3.7 of [5], it can be shown that the cyclicity of A_λ is also the smallest possible length of such a periodic behavior. Hence, the cyclicity of the matrix A can be seen as the minimal length of a periodic behavior in the sequence of the powers of A_λ.

Because of the existence of the integer N, mentioned in Theorem 3.9, it follows that there exists a smallest number $t(A)$ such that

$$A^{\otimes(k+\sigma)} = \lambda^{\otimes\sigma} \otimes A^{\otimes k}$$

for all $k \geq t(A)$. The number $t(A)$ will be called the *transient time* of A.

Theorem 3.9 gives a partial answer to the question about the limiting behavior of $x(k)$ defined in (3.1). Indeed, for any initial condition x_0, the sequence $x(k)$ will show a periodic behavior after at most $t(A)$ transitions as

$$\begin{aligned}
x(k+\sigma) &= A^{\otimes(k+\sigma)} \otimes x_0 \\
&= \lambda^{\otimes\sigma} \otimes A^{\otimes k} \otimes x_0 \\
&= \lambda^{\otimes\sigma} \otimes x(k),
\end{aligned}$$

for all $k \geq t(A)$. This periodic behavior is characterized through the eigenvalue and the cyclicity of A. If A has cyclicity one, then $x(k+1) = A \otimes x(k) = \lambda \otimes x(k)$ for $k \geq t(A)$ in the above equation. In words, for any initial vector x_0, $x(k)$ becomes an eigenvector of A for $k \geq t(A)$. Put differently, after $t(A)$ steps, $x(k)$ behaves like an eigenvector, and the effect of the initial value x_0 has died out. Hence, we use the name *transient time* for $t(A)$. In the stochastic literature, transient time is called *coupling time*.

Example 3.1.3   *Let*

$$A = \begin{pmatrix} -1 & 11 \\ 1 & \varepsilon \end{pmatrix}.$$

The powers of A are

$$A^{\otimes 2} = \begin{pmatrix} 12 & 10 \\ e & 12 \end{pmatrix}, \quad A^{\otimes 3} = \begin{pmatrix} 11 & 23 \\ 13 & 11 \end{pmatrix},$$

$$A^{\otimes 4} = \begin{pmatrix} 24 & 22 \\ 12 & 24 \end{pmatrix}, \quad A^{\otimes 5} = \begin{pmatrix} 23 & 35 \\ 25 & 23 \end{pmatrix}, \ldots,$$

and we conclude that

$$\begin{aligned}
A^{\otimes(k+2)} &= 12 \otimes A^{\otimes k} \\
&= 6^{\otimes 2} \otimes A^{\otimes k}, \qquad k \geq 2.
\end{aligned}$$

Hence, we obtain $\lambda(A) = 6$, $\sigma(A) = 2$, $\sigma_{g(A)} = 1$, and $t(A) = 2$. Moreover,

$$A_\lambda = \begin{pmatrix} -7 & 5 \\ -5 & \varepsilon \end{pmatrix},$$

and

$$A_\lambda^{\otimes 2} = \begin{pmatrix} e & -2 \\ -12 & e \end{pmatrix}, \quad A_\lambda^{\otimes 3} = \begin{pmatrix} -7 & 5 \\ -5 & -7 \end{pmatrix},$$

$$A_\lambda^{\otimes 4} = \begin{pmatrix} e & -2 \\ -12 & e \end{pmatrix}, \quad A_\lambda^{\otimes 5} = \begin{pmatrix} -7 & 5 \\ -5 & -7 \end{pmatrix}, \dots,$$

which gives

$$A_\lambda^* = A_\lambda^+ = \begin{pmatrix} e & 5 \\ -5 & e \end{pmatrix},$$

and, in accordance with Lemma 2.7, both columns of A_λ^* are eigenvectors of A. The fact that the columns are colinear is not a mere coincidence and will be elucidated in Chapter 4.

Even for matrices of small size, the transient time can be arbitrarily large. For example, matrix A with

$$A = \begin{pmatrix} -1 & -N \\ e & e \end{pmatrix},$$

where $N \in \mathbb{N}$ with $N \geq 2$, has transient time $t(A) = N$, while $\lambda(A) = e$ and $\sigma(A) = 1$.

3.2 THE CYCLE-TIME VECTOR: PRELIMINARY RESULTS

This and the following sections deal with the cycle-time vector of sequences $\{x(k) : k \in \mathbb{N}\}$ defined by $x(k+1) = A \otimes x(k)$ for all $k \geq 0$, with $A \in \mathbb{R}_{\max}^{n \times n}$ and $x(0) = x_0 \in \mathbb{R}_{\max}^n$, being the initial condition. In Sections 3.2.2 and 3.2.4 matrix A is assumed to be irreducible. In the other sections matrix A is just supposed to be square and regular.

3.2.1 Uniqueness of the cycle-time vector

As announced, first the dependency of the limit $\lim_{k \to \infty} x(k)/k$ on the initial condition will be investigated, assuming that the limit exists. The actual existence of this limit will be discussed later on. For this purpose, an appropriate norm will be introduced, the so-called l^∞-norm. The l^∞-norm of a vector $v \in \mathbb{R}^n$ is defined as the maximum of the absolute value of all entries of v and will be denoted by $||v||_\infty$. Hence, $||v||_\infty = \max_{i \in \underline{n}} |v_i|$ for every $v \in \mathbb{R}^n$, where $|.|$ denotes the absolute value. Note that the l^∞-norm of a vector in $\mathbb{R}_{\max}^n$ may be infinite. This happens if at least one of its components is equal to ε. However, when one considers only regular matrices and finite initial conditions, the asymptotic behavior can be expressed entirely in terms of vectors in $\mathbb{R}^n$ (i.e., in terms of finite vectors).

The following property plays a crucial role in proving that the asymptotic behavior is independent of the initial condition.

LEMMA 3.10 *Let $A \in \mathbb{R}_{\max}^{m \times n}$ be a regular (not necessarily square) matrix, then*

$$||(A \otimes u) - (A \otimes v)||_\infty \leq ||u - v||_\infty,$$

for any $u, v \in \mathbb{R}^n$.

Proof. Note that $A \otimes u, A \otimes v \in \mathbb{R}^m$, i.e., both vectors are finite. Set

$$\alpha \overset{\text{def}}{=} ||(A \otimes u) - (A \otimes v)||_\infty.$$

Then there is an $i_0 \in \underline{m}$ such that

$$\alpha = \left| \left[(A \otimes u) - (A \otimes v) \right]_{i_0} \right|.$$

Assume that $\alpha = [(A \otimes u) - (A \otimes v)]_{i_0} \geq 0$; then

$$\alpha = \max_{j \in \underline{n}} (a_{i_0 j} + u_j) - \max_{l \in \underline{n}} (a_{i_0 l} + v_l).$$

Hence, there is a $j_0 \in \underline{n}$ such that

$$\alpha = (a_{i_0 j_0} + u_{j_0}) - \max_{l \in \underline{n}} (a_{i_0 l} + v_l),$$

which is less than or equal to (take $l = j_0$)

$$(a_{i_0 j_0} + u_{j_0}) - (a_{i_0 j_0} + v_{j_0}) = u_{j_0} - v_{j_0},$$

implying that

$$\alpha \leq u_{j_0} - v_{j_0} \leq \max_{j \in \underline{n}} (u_j - v_j) \leq \max_{j \in \underline{n}} |u_j - v_j| = ||u - v||_\infty.$$

Hence, if $\alpha = [(A \otimes u) - (A \otimes v)]_{i_0} \geq 0$, then $\alpha \leq ||u - v||_\infty$. The same can be shown if $\alpha = [(A \otimes u) - (A \otimes v)]_{i_0} \leq 0$. Thus, the proof is completed. □

The property $||(A \otimes u) - (A \otimes v)||_\infty \leq ||u - v||_\infty$ for any $u, v \in \mathbb{R}^n$, mentioned in Lemma 3.10, is the so-called nonexpansiveness, in the l^∞-norm, of the mapping $u \in \mathbb{R}^n_{\max} \rightarrow A \otimes u \in \mathbb{R}^m_{\max}$. Repeated application of the lemma for a square regular matrix A yields that

$$||(A^{\otimes k} \otimes u) - (A^{\otimes k} \otimes v)||_\infty \leq ||u - v||_\infty, \qquad (3.5)$$

for any $u, v \in \mathbb{R}^n$ and all $k \geq 0$. In words, the l^∞-distance between $A^{\otimes k} \otimes u$ and $A^{\otimes k} \otimes v$ is bounded by $||u - v||_\infty$. The following theorem shows that non-expansiveness implies that the cycle-time vector, provided it exists for at least one initial vector, exists for any initial vector and is independent of the specific initial vector. Write $x(k; x_0)$ to express the dependency of $x(k)$ on its initial value, i.e., $x(k; x_0) = A^{\otimes k} \otimes x_0$.

THEOREM 3.11 *Consider the recurrence relation $x(k+1) = A \otimes x(k)$ for $k \geq 0$, with $A \in \mathbb{R}^{n \times n}_{\max}$ a square regular matrix and $x(0) = x_0$ as initial condition. If $x_0 \in \mathbb{R}^n$ is a particular initial condition such that the limit $\lim_{k \to \infty} x(k; x_0)/k$ exists, then this limit exists and has the same value for any initial condition $y_0 \in \mathbb{R}^n$.*

Proof. Assume that $x_0 \in \mathbb{R}^n$ is such that $\lim_{k \to \infty} x(k; x_0)/k = \eta$ with $\eta \in \mathbb{R}^n$. For any $y_0 \in \mathbb{R}^n$, nonexpansiveness implies

$$0 \leq \left\| \frac{x(k; y_0)}{k} - \frac{x(k; x_0)}{k} \right\|_\infty$$

$$\leq \frac{1}{k} || (A^{\otimes k} \otimes y_0) - (A^{\otimes k} \otimes x_0) ||_\infty$$

$$\leq \frac{1}{k} ||y_0 - x_0||_\infty.$$

Taking the limit as k tends to ∞ in the above row of inequalities yields

$$\lim_{k\to\infty}\left\|\frac{x(k;y_0)}{k}-\frac{x(k;x_0)}{k}\right\|_\infty=0.$$

Hence, as k tends to ∞ the l^∞-distance between $x(k;x_0)/k$ and $x(k;y_0)/k$ tends to zero, which implies that η is the cycle-time vector for any initial value y_0. Note that for the proof it is essential that all elements of x_0 and y_0 are finite.   □

3.2.2 Existence of the cycle-time vector for irreducible matrices

The consequence of Theorem 3.11 is that once the cycle-time vector exists, it is independent of the initial condition. Therefore, the next issue to be studied is the actual existence of this vector. In the special case where matrix A is irreducible, the existence of the cycle-time vector, actually the asymptotic growth rate, for a particular initial condition is obvious, as will be shown below.

LEMMA 3.12 *Consider the recurrence relation $x(k+1)=A\otimes x(k)$ for $k\geq 0$, with $A\in\mathbb{R}_{\max}^{n\times n}$ an irreducible matrix having eigenvalue $\lambda\in\mathbb{R}$. Then, for all $j\in\underline{n}$*

$$\lim_{k\to\infty}\frac{x_j(k;x_0)}{k}=\lambda$$

for any initial condition $x(0)=x_0\in\mathbb{R}^n$.

Proof. Let v be an eigenvector of A. Initializing the recurrence relation with $x_0=v$ gives

$$\lim_{k\to\infty}\frac{1}{k}x_j(k;x_0)=\lambda,$$

for all $j\in\underline{n}$; see (3.2). Since by Theorem 3.11 once the asymptotic growth rate exists, it is independent of x_0, and the proof is completed.   □

3.2.3 The generalized eigenmode for general matrices

In Lemma 3.12 the asymptotic growth rate for the recurrence relation $x(k+1)=A\otimes x(k)$ is characterized in the case where matrix A is irreducible. To give a similar characterization in the case where matrix A is not necessarily irreducible, the concept of a *generalized eigenmode* is introduced. In the following definition, the symbols $+$ and $\times$ stand for vector addition and scalar multiplication, respectively, in the conventional sense.

DEFINITION 3.13 *A pair of vectors $(\eta,v)\in\mathbb{R}^n\times\mathbb{R}^n$ is called a* generalized eigenmode *of the regular matrix A if for all $k\geq 0$*

$$A\otimes(k\times\eta+v)=(k+1)\times\eta+v.$$

The vector η in a generalized eigenmode will be shown to coincide with the cycle-time vector. We illustrate the above definition with the following example.

Example 3.2.1 *Consider*

$$A=\begin{pmatrix} a & b \\ \varepsilon & c \end{pmatrix},$$

with $a, c \in \mathbb{R}$ and $b \in \mathbb{R}_{\max}$. First, let $b = \varepsilon$. Then it is straightforward that

$$\left(\begin{pmatrix} a \\ c \end{pmatrix}, \begin{pmatrix} e \\ e \end{pmatrix} \right)$$

is a generalized eigenmode of A.

Now assume that $b \neq \varepsilon$. For $a \geq c$, it follows easily that

$$\left(\begin{pmatrix} a \\ c \end{pmatrix}, \begin{pmatrix} b - c \\ e \end{pmatrix} \right)$$

is a generalized eigenmode, and for $a \leq c$, a generalized eigenmode is given by

$$\left(\begin{pmatrix} c \\ c \end{pmatrix}, \begin{pmatrix} b - c \\ e \end{pmatrix} \right).$$

The notion of a generalized eigenmode can be seen as an extension of an eigenvalue/eigenvector pair, as will be illustrated in the following. For $\mu \in \mathbb{R}_{\max}$, let

$$\mathbf{u}[\mu] \overset{\text{def}}{=} \mu \otimes \mathbf{u}$$

denote the vector having value μ in each of its entries. Thus, if $\lambda \in \mathbb{R}$ and $v \in \mathbb{R}^n$ are such that $A \otimes v = \lambda \otimes v$, then $A \otimes \lambda^{\otimes k} \otimes v = \lambda^{\otimes(k+1)} \otimes v$ for all $k \geq 0$, implying that for all $k \geq 0$

$$A \otimes (k \times \mathbf{u}[\lambda] + v) = (k + 1) \times \mathbf{u}[\lambda] + v.$$

Hence, the cycle-time vector η can be seen as an extension of the notion of eigenvalue, whereas the vector v remains to play its role as eigenvector.

It follows that if a generalized eigenmode of a regular matrix exists, then the cycle-time vector exists and is unique. Indeed, assume that (η, v) is a generalized eigenmode of the regular matrix A and consider the recurrence relation $x(k+1) = A \otimes x(k)$ with $x_0 = v$. Then, by induction, it follows from the definition of a generalized eigenmode that $x(k) = k \times \eta + v$, so that the cycle-time vector satisfies $\lim_{k \to \infty} x(k)/k = \eta$, where it is crucial that the vector v is completely finite. From Theorem 3.11 it is known that this limit is independent of the initial condition, implying the uniqueness of the cycle-time vector.

Unlike the cycle-time vector, the second vector in a generalized eigenmode is not uniquely determined. Indeed, if (η, v) constitutes a generalized eigenmode of the regular matrix A, so does the pair $(\eta, \nu \otimes v)(= (\eta, \mathbf{u}[\nu] + v))$ for any $\nu \in \mathbb{R}$.

3.2.4 Preliminary results on inhomogeneous recurrence relations

In order to have a better understanding of the limiting behavior of reducible matrices, we will study a natural extension of the recurrence relation $x(k+1) = A \otimes x(k)$ in which A is an irreducible matrix. To that end, consider the recurrence relation

$$x(k + 1) = A \otimes x(k) \oplus \bigoplus_{j=1}^{m} B_j \otimes u_j(k), \tag{3.6}$$

where A is an $n \times n$ matrix over $\mathbb{R}_{\max}$ and $B_1, \ldots, B_m$ are matrices over $\mathbb{R}_{\max}$ of suitable sizes; that is, for each $j \in \underline{m}$, matrix B_j is an $n \times m_j$ matrix over $\mathbb{R}_{\max}$ for some appropriate $m_j \geq 1$, and $u_j(k)$ denotes a vector (!) with m_j elements. The latter is in contrast with the notation so far, where $u_j(k)$ is used to indicate the jth element of $u(k)$. The reader should be warned that for the time being $u_j(k)$ will denote a vector of suitable size for any $j \in \underline{m}$.

Next assume the following:

- A is irreducible, so that the eigenvalue of A, denoted by $\lambda = \lambda(A)$, exists.

- Each of the matrices $B_1, \ldots, B_m$ contains at least one finite element, i.e., $B_j \neq \mathcal{E}$ for all $j \in \underline{m}$.

- For $j \in \underline{m}$, each of the sequences $u_j(k), k \geq 0$, is of the form

$$u_j(k) = w_j \otimes \tau_j^{\otimes k}, \qquad k \geq 0,$$

for some vector $w_j \in \mathbb{R}^{m_j}$ and scalar $\tau_j \in \mathbb{R}$.

Denote $\tau = \bigoplus_{j \in \underline{m}} \tau_j$, i.e., $\tau = \max\{\tau_1, \ldots, \tau_m\}$. Now it is claimed that there exists an integer $K \geq 0$ and a vector $v \in \mathbb{R}^n$ such that the sequence defined by

$$x(k) = v \otimes \mu^{\otimes k}, \quad \text{with} \quad \mu = \lambda \oplus \tau,$$

satisfies the recurrence relation (3.6) for all $k \geq K$.

To prove the claim, two cases will be distinguished, namely, $\lambda > \tau$ and $\lambda \leq \tau$.

Case $\lambda > \tau$. Take v to be an eigenvector of matrix A corresponding to eigenvalue λ, and recall that v is finite by the irreducibility of A, i.e., $v \in \mathbb{R}^n$; see Lemma 2.8. Further, choose v such that $v \otimes \lambda > \bigoplus_{j=1}^{m} B_j \otimes w_j$. The latter inequality can always be satisfied in combination with $A \otimes v = \lambda \otimes v$. Indeed, if $A \otimes v = \lambda \otimes v$, but not $\lambda \otimes v > \bigoplus_{j=1}^{m} B_j \otimes w_j$, then replace v by $v \otimes \rho$ with

$$\rho > \left(\bigoplus_{j=1}^{m} B_j \otimes w_j)^{\text{top}} + (-(\lambda \otimes v)) \right)^{\text{top}}.$$

In the latter, the notation γ^{top} denotes the maximal element of vector γ. Then, with $\mu = \lambda > \tau_j$ for all $j \in \underline{m}$, it follows that for all $k \geq 0$

$$\mu \otimes v \otimes \mu^{\otimes k} = A \otimes v \otimes \mu^{\otimes k} > \bigoplus_{j=1}^{m} B_j \otimes w_j \otimes \mu^{\otimes k} \geq \bigoplus_{j=1}^{m} B_j \otimes w_j \otimes \tau_j^{\otimes k}.$$

Hence,

$$v \otimes \mu^{\otimes(k+1)} = A \otimes v \otimes \mu^{\otimes k} > \bigoplus_{j=1}^{m} B_j \otimes w_j \otimes \tau_j^{\otimes k}$$

for all $k \geq 0$, so that recurrence relation (3.6) is fulfilled for all $k \geq 0$, with $x(k) = v \otimes \mu^{\otimes k}$ and $u_j(k) = w_j \otimes \tau_j^{\otimes k}$ for $j \in \underline{m}$.

Case $\lambda \leq \tau$. Recall that $\tau = \max\{\tau_1, \ldots, \tau_m\}$. Assume without loss of generality that $\tau = \tau_j$ for $j = 1, \ldots, r$ and that $\tau > \tau_j$ for $j = r+1, \ldots, m$, with $1 < r \leq m$; that is, the maximum in $\max\{\tau_1, \ldots, \tau_m\}$ is attained by the first

r τ's. The latter can always be accomplished by a renumbering of the sequences $u_j(k)$, $j \in \underline{m}$. Now take vector v to be a solution of

$$v = A_\tau \otimes v \oplus \bigoplus_{j=1}^{r} (B_j)_\tau \otimes w_j, \tag{3.7}$$

where A_τ denotes the matrix obtained from matrix A by subtracting τ from all of its finite elements and similarly for $(B_j)_\tau$, for $j \in \underline{m}$. Because $\lambda \leq \tau$, the communication graph of A_τ only contains circuits with a nonpositive weight. Therefore, it follows from Theorem 2.10 that a solution v of (3.7) is given by

$$v = (A_\tau)^* \otimes \left(\bigoplus_{j=1}^{r} (B_j)_\tau \otimes w_j \right).$$

Because A (and thus A_τ) is irreducible, matrix $(A_\tau)^*$ is completely finite. Further, since $\bigoplus_{j=1}^{r} (B_j)_\tau \otimes w_j$ contains at least one finite element, it follows that v is finite (i.e., $v \in \mathbb{R}^n$). Note that by adding τ to both sides of (3.7), it follows that v satisfies

$$v \otimes \tau = A \otimes v \oplus \bigoplus_{j=1}^{r} B_j \otimes w_j.$$

Then, with $\mu = \tau = \tau_j$ for $j = 1, \ldots, r$, it follows for all $k \geq 0$ that

$$v \otimes \mu^{\otimes(k+1)} = A \otimes v \otimes \mu^{\otimes k} \oplus \bigoplus_{j=1}^{r} B_j \otimes w_j \otimes \tau_j^{\otimes k},$$

leading to the inequality

$$v \otimes \mu^{\otimes(k+1)} \leq A \otimes v \otimes \mu^{\otimes k} \oplus \bigoplus_{j=1}^{m} B_j \otimes w_j \otimes \tau_j^{\otimes k}. \tag{3.8}$$

However, since $\mu > \tau_j$ for $j = r+1, \ldots, m$, there exists an integer $K \geq 0$ such that for all $k \geq K$

$$v \otimes \mu^{\otimes(k+1)} \geq \bigoplus_{j=r+1}^{m} B_j \otimes w_j \otimes \tau_j^{\otimes k}.$$

Hence, for all $k \geq K$ the inequality in (3.8) in fact is an equality, so that the recurrence relation (3.6) is fulfilled for all $k \geq K$, with $x(k) = v \otimes \mu^{\otimes k}$ and $u_j(k) = w_j \otimes \tau_j^{\otimes k}$, for $j \in \underline{m}$.

As both cases are treated, the following theorem has been proved. In fact, if the above is repeated for $A = \varepsilon$, it follows easily that the following theorem holds in the case where $A = \varepsilon$.

THEOREM 3.14 *Consider the recurrence relation given by (3.6) and assume the following:*

- *A is irreducible with eigenvalue $\lambda = \lambda(A)$, or $A = \varepsilon$ with $\lambda = \varepsilon$;*

- $B_j \neq \mathcal{E}$ for all $j \in \underline{m}$;

- $u_j(k) = w_j \otimes \tau_j^{\otimes k}$, $k \geq 0$, with $\tau_j \in \mathbb{R}$ and w_j a finite vector of suitable size for all $j \in \underline{m}$.

Denote $\tau = \bigoplus_{j \in \underline{m}} \tau_j$. Then there exists an integer $K \geq 0$ and a vector $v \in \mathbb{R}^n$ such that the sequence defined by

$$x(k) = v \otimes \mu^{\otimes k}, \qquad \text{with} \qquad \mu = \lambda \oplus \tau,$$

satisfies (3.6) for all $k \geq K$.

Note that in Theorem 3.14 recurrence relation (3.6) is satisfied for k larger than or equal to some integer $K \geq 0$. However, in the case where it is possible to reinitialize the sequences $u_j(k) = w_j \otimes \tau_j^{\otimes k}$, $k \geq 0$, by redefining the vectors w_j for $j \in \underline{m}$, it is possible that recurrence relation (3.6) is fulfilled for all $k \geq 0$, i.e., that K is actually zero. Indeed, in such a case, redefine

$$v := v \otimes \mu^{\otimes K}, \qquad w_j := w_j \otimes \tau_j^{\otimes K}, \qquad j \in \underline{m},$$

where v, μ, and K come from Theorem 3.14. Then the sequences

$$x(k) = v \otimes \mu^{\otimes k}, \qquad u_j(k) = w_j \otimes \tau_j^{\otimes k}, \qquad j \in \underline{m},$$

satisfy recurrence relation (3.6) for $k \geq 0$.

3.3 THE CYCLE-TIME VECTOR: GENERAL RESULTS

In this section the existence of a generalized eigenmode for a square reducible matrix $A \in \mathbb{R}_{\max}^{n \times n}$ will be proved. In doing so, also the existence of the cycle-time vector of a square reducible matrix is proved. Further, we will derive explicit expressions for the value of the elements of the cycle-time vector.

3.3.1 Existence of the cycle-time vector for reducible matrices

Consider the recurrence relation

$$x(k+1) = A \otimes x(k), \tag{3.9}$$

with A reducible. Recall that by renumbering the nodes in the communication graph $\mathcal{G}(A)$, matrix A can be brought into a block upper triangular form, called a *normal form* of A, given by

$$
\begin{pmatrix}
A_{11} & A_{12} & \cdots & \cdots & A_{1q} \\
\mathcal{E} & A_{22} & \cdots & \cdots & A_{2q} \\
\mathcal{E} & \mathcal{E} & A_{33} & & \vdots \\
\vdots & \vdots & \ddots & \ddots & \vdots \\
\mathcal{E} & \mathcal{E} & \cdots & \mathcal{E} & A_{qq}
\end{pmatrix},
$$

with, the conditions in which for $i \in \underline{q}$,

- either A_{ii} is an irreducible matrix, so that $\lambda_i = \lambda(A_{ii})$ exists, or

- $A_{ii} = \varepsilon$, in which case $\lambda_i = \varepsilon$.

Let the vector $x(k)$ be partitioned according to the above normal form of A as

$$\begin{pmatrix} x_1(k) \\ x_2(k) \\ \vdots \\ x_q(k) \end{pmatrix},$$

where $x_i(k)$ for $i \in \underline{q}$ denotes a vector (!) of suitable size. Again this is in contrast with the notation so far, where $x_i(k)$ is used to indicate the ith element of $x(k)$. Hence, the reader should be aware that for the time being $x_i(k)$ for $i \in \underline{q}$ in principle is a vector.

It follows now that recurrence relation (3.9) can be written as

$$x_i(k+1) = A_{ii} \otimes x_i(k) \oplus \bigoplus_{j=i+1}^{q} A_{ij} \otimes x_j(k), \quad i \in \underline{q}. \tag{3.10}$$

In particular, it follows for $i = q$ that

$$x_q(k+1) = A_{qq} \otimes x_q(k).$$

Since it is assumed throughout that matrix A is regular (i.e., contains at least one finite element in each row), it follows that also A_{qq} is regular. So $A_{qq} \neq \mathcal{E}$, implying that the corresponding m.s.c.s. has a nonempty arc set and, consequently, A_{qq} is irreducible. Hence, there exist a finite vector v_q and a scalar $\xi_q \in \mathbb{R}$ such that the sequence

$$x_q(k) = v_q \otimes \xi_q^{\otimes k}$$

satisfies $x_q(k+1) = A_{qq} \otimes x_q(k)$ for all $k \geq 0$. Indeed, simply take for v_q an eigenvector of A_{qq} corresponding to the eigenvalue $\lambda_q = \lambda(A_{qq})$, and set $\xi_q = \lambda_q$.

The case for general $i \in \underline{q}$ is treated in the following theorem, which will be proved by means of induction with respect to i, going from q back to 1.

THEOREM 3.15 *Consider the recurrence relations given in (3.10). Assume that A_{qq} is irreducible and that for $i \in \underline{q-1}$ either A_{ii} is an irreducible matrix or $A_{ii} = \varepsilon$. Then there exist finite vectors $v_1, v_2, \ldots, v_q$ of suitable sizes and scalars $\xi_1, \xi_2, \ldots, \xi_q \in \mathbb{R}$ such that the sequences*

$$x_i(k) = v_i \otimes \xi_i^{\otimes k}, \qquad i \in \underline{q},$$

satisfy (3.10) for all $k \geq 0$. The scalars $\xi_1, \xi_2, \ldots, \xi_q$ are determined by

$$\xi_i = \bigoplus_{j \in \mathcal{H}_i} \xi_j \oplus \lambda_i,$$

where $\mathcal{H}_i = \{j \in \underline{q} : j > i, A_{ij} \neq \mathcal{E}\}$.

Proof. For $i = q$, the result is immediate. Next, assume that the result is true for some $l + 1$, with $1 < l + 1 \leq q$. Hence, there are finite vectors $v_{l+1}, \ldots, v_q$ of suitable sizes and scalars $\xi_{l+1}, \ldots, \xi_q \in \mathbb{R}$ such that the sequences

$$x_i(k) = v_i \otimes \xi_i^{\otimes k}, \qquad l + 1 \leq i \leq q,$$

satisfy

$$x_i(k + 1) = A_{ii} \otimes x_i(k) \oplus \bigoplus_{j=i+1}^{q} A_{ij} \otimes x_j(k), \qquad l + 1 \leq i \leq q,$$

for all $k \geq 0$. Next, consider

$$x_l(k + 1) = A_{ll} \otimes x_l(k) \oplus \bigoplus_{j=l+1}^{q} A_{lj} \otimes x_j(k). \qquad (3.11)$$

Recall that either A_{ll} is a square irreducible matrix of suitable size or A_{ll} is a 1×1 matrix (i.e., a scalar) equal to ε. Further, note that

$$\bigoplus_{j=l+1}^{q} A_{lj} \otimes x_j(k) = \bigoplus_{j \in \mathcal{H}_l} A_{lj} \otimes x_j(k),$$

with $\mathcal{H}_l = \{j \in \underline{q} : j > l, A_{lj} \neq \mathcal{E}\}$. Hence, (3.11) can be written as

$$x_l(k + 1) = A_{ll} \otimes x_l(k) \oplus \bigoplus_{j \in \mathcal{H}_l} A_{lj} \otimes x_j(k). \qquad (3.12)$$

By this all the assumptions in Theorem 3.14 with respect to the recurrence relation in (3.12) are satisfied. It therefore follows from Theorem 3.14 that there exists an integer $K_l \geq 0$ and a finite vector/scalar v_l such that the sequence

$$x_l(k) = v_l \otimes \xi_l^{\otimes k}$$

satisfies the recurrence relation (3.12) or, equivalently, (3.11) for all $k \geq K_l$, where

$$\xi_l = \bigoplus_{j \in \mathcal{H}_l} \xi_j \oplus \lambda_l .$$

All the sequences $x_i(k), l \leq i \leq q$, can be reinitialized by redefining

$$v_i := v_i \otimes \xi_i^{\otimes K_l}, \qquad l \leq i \leq q.$$

It then follows that the (new) sequences

$$x_i(k) = v_i \otimes \xi_i^{\otimes k}, \qquad l \leq i \leq q,$$

satisfy

$$x_i(k + 1) = A_{ii} \otimes x_i(k) \oplus \bigoplus_{j=i+1}^{q} A_{ij} \otimes x_j(k), \qquad l \leq i \leq q,$$

for all $k \geq 0$. Hence, the result is also true for l, with $1 \leq l < q$. This concludes the induction step and also the proof of the theorem. $\square$

The following corollary is an immediate consequence of Theorem 3.15.

COROLLARY 3.16 *Consider the recurrence relation given in (3.9), and assume that A is regular. Then there exist finite vectors $\eta, v \in \mathbb{R}^n$ such that the sequence*

$$x(k) = k \times \eta + v$$

satisfies (3.9) for all $k \geq 0$.

Proof. First, by permuting/renumbering, transform the recurrence relation (3.9) into the recurrence relations (3.10). Note that due to the regularity of A the diagonal matrices in the normal form of A satisfy the conditions of Theorem 3.15. Observe that the sequences $x_i(k) = v_i \otimes \xi_i^{\otimes k}$, resulting from Theorem 3.15, also can be written as

$$x_i(k) = k \times \mathbf{u}[\xi_i] + v_i.$$

Then stacking the $\mathbf{u}[\xi_i]$'s on top of each other, stacking the v_i's on top of each other, and permuting/renumbering backwards, two vectors remain, say, η and v, such that the sequence

$$x(k) = k \times \eta + v$$

can easily be shown to satisfy (3.9) for all $k \geq 0$. $\qquad\square$

Corollary 3.16 shows that any regular matrix $A \in \mathbb{R}_{\max}^{n \times n}$ has a generalized eigenmode. Consequently, the cycle-time vector exists for any regular matrix $A \in \mathbb{R}_{\max}^{n \times n}$.

Note that the cycle-time vector of matrix A in the normal form, as used above, is given by

$$\eta = \left(\mathbf{u}^\top[\xi_1], \mathbf{u}^\top[\xi_2], \ldots, \mathbf{u}^\top[\xi_q]\right)^\top.$$

For matrices A not in normal form, the cycle-time vector is a reshuffled version of the cycle-time vector corresponding to a normal form of A.

3.3.2 Expressions for the elements of the cycle-time vector

Consider again recurrence relation (3.9), and write it in the form of recurrence relations (3.10) by bringing A into normal form.

Assume, as in Section 2.1, that matrix A_{ss} in (3.10) corresponds to the restriction of matrix A to the m.s.c.s. denoted by $[i_s]$. Also note that $A_{sr} \neq \mathcal{E}$ in (3.10) if there exists an arc in $\mathcal{G}(A)$ from a node in m.s.c.s. $[i_r]$ to a node in m.s.c.s. $[i_s]$.

According to Theorem 3.15 the cycle-time vector corresponding to the m.s.c.s. $[i_s]$ is given by $\mathbf{u}[\xi_s]$. This follows simply from the solution for $x_s(k)$, given in Theorem 3.15, written as $x_s(k) = k \times \mathbf{u}[\xi_s] + v_s$. Hence, asymptotically speaking, all components of $x(k)$ corresponding to the nodes in the m.s.c.s. $[i_s]$ grow with the same speed, namely, ξ_s, where these scalars are determined by

$$\xi_s = \bigoplus_{r \in \mathcal{H}_s} \xi_r \oplus \lambda_s, \tag{3.13}$$

with $\mathcal{H}_s = \{r \in \underline{q} : r > s, A_{sr} \neq \mathcal{E}\}$. By the above equations the scalars ξ_s, for $s \in \underline{q}$, are determined implicitly. To have explicit expressions for these scalars, consider the following $q \times q$ matrix $\widetilde{A}$ defined by

$$[\widetilde{A}]_{sr} = \begin{cases} 0 & r > s, A_{sr} \neq \mathcal{E} \text{ (i.e., } r \in \mathcal{H}_s), \\ \varepsilon & \text{otherwise.} \end{cases}$$

The equation in (3.13) can then be written as

$$\xi = \widetilde{A} \otimes \xi \oplus \lambda, \qquad (3.14)$$

where

$$\xi = (\xi_1, \xi_2, \ldots, \xi_q)^\top, \qquad \lambda = (\lambda_1, \lambda_2, \ldots, \lambda_q)^\top.$$

In the communication graph $\mathcal{G}(\widetilde{A})$ of matrix $\widetilde{A}$, identify each node with the m.s.c.s. it represents in the communication graph $\mathcal{G}(A)$ of matrix A. Hence, $\mathcal{G}(\widetilde{A})$ has node set $\{[i_1], [i_2], \ldots, [i_q]\}$. Further, note that $\mathcal{G}(\widetilde{A})$ has arcs with weight zero and that $\mathcal{G}(\widetilde{A})$ coincides with the reduced graph of A, introduced in Section 2.1.

Because matrix $\widetilde{A}$ is strictly upper triangular, its communication graph $\mathcal{G}(\widetilde{A})$ does not contain any circuits. Therefore, according to Theorem 2.10, the (unique) solution of (3.14) can be written as

$$\xi = (\widetilde{A})^* \otimes \lambda.$$

Recall that $[(\widetilde{A})^*]_{sr}$ is the maximal weight of a path in $\mathcal{G}(\widetilde{A})$ from node $[i_r]$ to node $[i_s]$ and that $[(\widetilde{A})^*]_{sr} = \varepsilon$ if no such path exists. Since all arcs have weight zero, every path has weight zero. Hence, it follows from $\xi = (\widetilde{A})^* \otimes \lambda$ that for all $s \in \underline{q}$ component ξ_s of vector ξ satisfies

$$\xi_s = \max\{\lambda_r | \text{there is a path in } \mathcal{G}(\widetilde{A}) \text{ from node } [i_r] \text{ to node } [i_s]\}.$$

If there is a path in $\mathcal{G}(\widetilde{A})$ from node $[i_r]$ to node $[i_s]$, then there is a path in $\mathcal{G}(A)$ from any node $i \in [i_r]$ to any node $j \in [i_s]$ and conversely. From Theorem 3.15 it follows that all nodes in one m.s.c.s. have the same asymptotic growth rate. Hence, their asymptotic growth rate can be identified with the asymptotic growth rate of the m.s.c.s. they belong to. Recall that $[j]$ stands for the m.s.c.s. that node j belongs to. Hence, $\xi_{[j]}$ denotes the asymptotic growth rate corresponding to m.s.c.s. $[j]$. Similarly, $\lambda_{[i]}$ denotes the eigenvalue corresponding to m.s.c.s. $[i]$, i.e., the eigenvalue of $A_{[i][i]}$. Finally, recall that $\pi^*(j)$ stands for the set of nodes i from which there is a path in $\mathcal{G}(A)$ to node j, including node j itself. Then the next theorem follows from the previous discussion.

THEOREM 3.17 *Consider the recurrence relation* $x(k+1) = A \otimes x(k)$ *for* $k \geq 0$, *with a square regular matrix* $A \in \mathbb{R}_{\max}^{n \times n}$ *and an initial condition* $x(0) = x_0$. *Let* $\xi = \lim_{k \to \infty} x(k; x_0)/k$ *be the cycle-time vector of* A.

1. *For all* $j \in \underline{n}$,

$$\xi_{[j]} = \bigoplus_{i \in \pi^*(j)} \lambda_{[i]}.$$

2. *For all* $j \in \underline{n}$ *and any* $x_0 \in \mathbb{R}^n$,

$$\lim_{k \to \infty} \frac{1}{k} x_j(k; x_0) = \bigoplus_{i \in \pi^*(j)} \lambda_{[i]}.$$

Example 3.3.1 *Consider A as given in Example 2.1.3. The communication graph of A has two nontrivial m.s.c.s.'s, given by the node sets $\{1, 2, 3, 4\}$ and $\{5, 6, 7\}$. The associated arc sets follow directly from the graph in Figure 2.2. The other m.s.c.s.'s in the graph each consist of a single node with no arcs attached. Inspecting the graph in Figure 2.2, it immediately follows that*

$$\lambda_{[1]} = \lambda_{[2]} = \lambda_{[3]} = \lambda_{[4]} = \frac{1}{2}, \ \lambda_{[5]} = \lambda_{[6]} = \lambda_{[7]} = \frac{4}{3}, \ \lambda_{[8]} = \lambda_{[9]} = \lambda_{[10]} = \varepsilon.$$

Applying Theorem 3.17 then gives (see also Figure 2.3)

$$\lim_{k \to \infty} \frac{x_j(k)}{k} = \lambda_{[2]} = \tfrac{1}{2}, \qquad j \in \{1, 2, 3, 4\},$$

$$\lim_{k \to \infty} \frac{x_j(k)}{k} = \max\left(\lambda_{[2]}, \lambda_{[5]}\right) = \tfrac{4}{3}, \qquad j \in \{5, 6, 7\},$$

$$\lim_{k \to \infty} \frac{x_8(k)}{k} = \max\left(\lambda_{[8]}, \lambda_{[2]}, \lambda_{[5]}\right) = \tfrac{4}{3},$$

$$\lim_{k \to \infty} \frac{x_9(k)}{k} = \max\left(\lambda_{[9]}, \lambda_{[2]}, \lambda_{[5]}\right) = \tfrac{4}{3},$$

$$\lim_{k \to \infty} \frac{x_{10}(k)}{k} = \max\left(\lambda_{[10]}, \lambda_{[2]}\right) = \tfrac{1}{2},$$

for any finite initial value x_0.

3.4 A SUNFLOWER BOUQUET

In this last section a special class of matrices is considered that will play an important role in Chapter 6.

DEFINITION 3.18 *A matrix $A \in \mathbb{R}_{\max}^{n \times n}$ is called a sunflower matrix if*

- *its communication graph $\mathcal{G}(A)$ consists of precisely one circuit and possibly a number of paths and*

- *each node of $\mathcal{G}(A)$, being part of the circuit or of a possible path, has one incoming arc.*

The communication graph of a sunflower matrix will be called a sunflower graph.

Another way of defining such a sunflower graph is to say that a node not already belonging to the unique circuit is connected via a unique path (in backward direction) to this circuit.

A sunflower matrix is a special kind of regular matrix because it contains in every row precisely one finite entry. Indeed, in the communication graph of a sunflower matrix every node has precisely one predecessor. Further, note that in the graph of a sunflower matrix the circuit is the only strongly connected part and that every node can be reached from this circuit. Hence, if the (maximum) circuit mean is λ, then according to Theorem 3.17 it follows that the asymptotic growth rate of each component equals λ. The cycle-time vector of the matrix is thus given by $\mathbf{u}[\lambda]$, where λ is the circuit mean of the only circuit. To compute the second vector v of a generalized eigenmode of a sunflower matrix A, let the matrix be specified by its n finite entries $a_{i\pi(i)}$ for $i \in \underline{n}$. Then $v \in \mathbb{R}^n$, together with $\mathbf{u}[\lambda]$, should satisfy

$$A \otimes (k \times \mathbf{u}[\lambda] + v) = (k+1) \times \mathbf{u}[\lambda] + v,$$

for all $k \geq 0$ or in more detail, $a_{i\pi(i)} + k \times \lambda + v_{\pi(i)} = (k+1) \times \lambda + v_i$, for all $i \in \underline{n}$ and for all $k \geq 0$. This can be simplified to

$$a_{i\pi(i)} - \lambda + v_{\pi(i)} = v_i,$$

for all $i \in \underline{n}$. Now let i_0 be a node of the circuit and set $v_{i_0} = 0$. Then the components of v corresponding to the other nodes of the circuit can be obtained by going backward around the circuit using $v_{\pi(i)} = v_i - a_{i\pi(i)} + \lambda$. The components of v corresponding to any of the paths can be obtained by going forward along the paths using $v_i = a_{i\pi(i)} - \lambda + v_{\pi(i)}$. Clearly, the latter computations can be done very efficiently and require a number of operations that is linear in n. The above discussion is summarized in the following lemma.

LEMMA 3.19 *Let A be a sunflower matrix. Then $(\mathbf{u}[\lambda], v)$ is a generalized eigenmode $(\mathbf{u}[\lambda], v)$ of A, where λ is the circuit mean of the unique circuit in $\mathcal{G}(A)$ and v is recursively given through*

$$a_{i\pi(i)} - \lambda + v_{\pi(i)} = v_i,$$

for all $i \in \underline{n}$, with $v_{i_0} = 0$ for some initial node i_0 on the circuit.

If the graph of a sunflower matrix A contains nodes that do not belong to the circuit, then A is reducible. Lemma 3.19 can thus be phrased by saying that a sunflower matrix possesses a unique eigenvalue, which is an extension of the result in Theorem 2.9 to a subclass of reducible matrices.

A *bouquet matrix* is a regular matrix whose communication graph consists of a number of disjoint sunflower graphs. It can be seen easily that a bouquet matrix has precisely one finite entry in each row, implying that each node of its communication graph has precisely one predecessor. See Figure 3.2 for an example of the communication graph of a bouquet matrix. Since the communication graph of a bouquet matrix A consists of $r \geq 1$ disjoint graphs, matrix A can (after possible relabeling of the nodes in $\mathcal{G}(A)$) be written in the following block diagonal form:

$$A = \begin{pmatrix} A_1 & \mathcal{E} & \cdots & & & \mathcal{E} \\ \mathcal{E} & A_2 & \mathcal{E} & & & \vdots \\ \vdots & \mathcal{E} & \ddots & \ddots & & \\ & & \ddots & A_{r-1} & \mathcal{E} \\ \mathcal{E} & \cdots & & \mathcal{E} & A_r \end{pmatrix},$$

where each of the block matrices A_i, $i \in \underline{r}$, is a sunflower matrix. Note that the above form should not be confused with the normal form of matrix A, in which

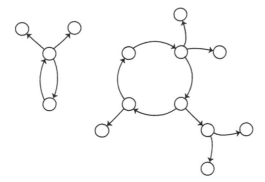

Figure 3.2: Communication graph of a bouquet matrix (consisting of two sunflowers).

each of the diagonal block matrices is either irreducible or equal to ε. In general, sunflower matrices are neither irreducible nor equal to ε. Clearly, Lemma 3.19 can be applied to the individual block matrices A_i, $i \in \underline{r}$, yielding generalized eigenmodes $(\mathbf{u}[\lambda_i], v_i)$, $i \in \underline{r}$, respectively. Let

$$v = (v_1^\top, v_2^\top, \ldots, v_r^\top)^\top \tag{3.15}$$

and

$$\eta = (\mathbf{u}^\top[\lambda_1], \mathbf{u}^\top[\lambda_2], \ldots, \mathbf{u}^\top[\lambda_r])^\top. \tag{3.16}$$

Then, it is easily seen that (η, v) is a generalized eigenmode of bouquet matrix A.

With the above method the generalized eigenmode of a bouquet matrix can in principle be computed. In Section 6.1.1 the latter computations will be presented in a more algorithmic form.

3.5 EXERCISES

1. Let $A \in \mathbb{R}_{\max}^{n \times n}$ be a regular square matrix. Recall that $|| \cdot ||_\infty$ denotes the l^∞-norm of a vector. Show the following:

    (a) $||A \otimes u - A \otimes v||_\infty \leq ||u - v||_\infty$ for all $u, v \in \mathbb{R}^n$;

    (b) $||A^{\otimes k} \otimes u - u||_\infty \leq k \times ||A \otimes u - u||_\infty$ for all $k \in \mathbb{N}$ and $u \in \mathbb{R}^n$;

    (c) if η denotes the cycle-time vector of A, then $||\eta||_\infty \leq ||A \otimes u - u||_\infty$ for all $u \in \mathbb{R}^n$, implying that $||\eta||_\infty \leq \min_{u \in \mathbb{R}^n} ||A \otimes u - u||_\infty$.

2. Let $A \in \mathbb{R}_{\max}^{n \times n}$ be irreducible, and let e_j be the jth base vector $j \in \underline{n}$. Then for all $i \in \underline{n}$, there exists an $l \in \underline{n}$ such that $[A^{\otimes l} \otimes e_j]_i \neq \varepsilon$. This implies that for all $i \in \underline{n}$ there exists an $l \in \underline{n}$, such that $[A^{\otimes l} \otimes v]_i \neq \varepsilon$ for any for $v \neq \mathbf{u}[\varepsilon]$.

3. Let $A \in \mathbb{R}_{\max}^{n \times n}$ be an irreducible matrix having eigenvalue λ. Then $\lim_{k \to \infty} A^{\otimes k}/k$ is an $n \times n$ matrix with λ in each of its entries.

4. Let $A \in \mathbb{R}_{\max}^{n \times n}$ and $B \in \mathbb{R}_{\max}^{n \times m}$ be given matrices. Consider the inhomogeneous recurrence relation $x(k+1) = A \otimes x(k) \oplus B \otimes u(k)$. Let $x(0)$ and $u(k)$ for all $k \in \mathbb{N}$ be given. Then prove that for all $k \in \mathbb{N}$

$$x(k) = A^{\otimes k} \otimes x(0) \oplus \bigoplus_{l=0}^{k-1} A^{\otimes l} \otimes B \otimes u(k-1-l).$$

5. Show that the relation defined in (3.3) on the node set $\mathcal{N}(A)$ of a communication graph $\mathcal{G}(A)$ with cyclicity $\sigma_{\mathcal{G}(A)}$ is an equivalence relation with equivalence classes defined in (3.4).

6. Consider the matrix

$$A = \begin{pmatrix} \varepsilon & \varepsilon & 2 & \varepsilon \\ \varepsilon & \varepsilon & \varepsilon & 1 \\ 0 & 4 & \varepsilon & 0 \\ \varepsilon & \varepsilon & 1 & \varepsilon \end{pmatrix},$$

and sketch its communication graph $\mathcal{G}(A)$ and its critical graph $\mathcal{G}^c(A)$. From the graphs, determine $\sigma_{\mathcal{G}(A)}, \sigma(A)$, and $\lambda(A)$. Compute $t(A)$ by straightforward (but tedious) multiplication. $\left(\text{Answer: } \sigma_{\mathcal{G}(A)} = 1, \sigma(A) = 3, \lambda(A) = 2, t(A) = 6.\right)$

7. By looking at the matrix

$$A = \begin{pmatrix} 1 & -6 \\ 0 & 2 \end{pmatrix}$$

and some of its first powers, determine $\lambda(A)$, and $\sigma(A)$ and estimate $t(A)$. $\left(\text{Answer: } \lambda(A) = 2, \sigma(A) = 1, t(A) = 10.\right)$

8. Determine a generalized eigenmode of the matrices

$$A = \begin{pmatrix} \varepsilon & 3 & \varepsilon \\ 1 & \varepsilon & 1 \\ \varepsilon & \varepsilon & 4 \end{pmatrix}, \quad A = \begin{pmatrix} 4 & 1 & \varepsilon \\ \varepsilon & \varepsilon & 3 \\ \varepsilon & 1 & \varepsilon \end{pmatrix}.$$

$\left(\text{Answer: } (\eta, v) = ((4, 4, 4)^\top, (0, 1, 4)^\top), (\eta, v) = ((4, 2, 2)^\top, (1, 1, 0)^\top).\right)$

9. Check that the following matrix is a sunflower matrix, and compute a generalized eigenmode as explained in Section 3.4:

$$A = \begin{pmatrix} \varepsilon & 2 & \varepsilon & \varepsilon & \varepsilon & \varepsilon & \varepsilon & \varepsilon \\ \varepsilon & \varepsilon & 7 & \varepsilon & \varepsilon & \varepsilon & \varepsilon & \varepsilon \\ 3 & \varepsilon & \varepsilon & \varepsilon & \varepsilon & \varepsilon & \varepsilon & \varepsilon \\ \varepsilon & \varepsilon & 4 & \varepsilon & \varepsilon & \varepsilon & \varepsilon & \varepsilon \\ \varepsilon & \varepsilon & 5 & \varepsilon & \varepsilon & \varepsilon & \varepsilon & \varepsilon \\ \varepsilon & 2 & \varepsilon & \varepsilon & \varepsilon & \varepsilon & \varepsilon & \varepsilon \\ \varepsilon & 6 & \varepsilon & \varepsilon & \varepsilon & \varepsilon & \varepsilon & \varepsilon \\ \varepsilon & \varepsilon & \varepsilon & \varepsilon & \varepsilon & \varepsilon & 7 & \varepsilon \end{pmatrix}.$$

$\left(\text{Answer: } (\eta, v) = ((4, 4, 4, 4, 4, 4, 4, 4)^\top, (0, 2, -1, -1, 0, 0, 4, 7)^\top).\right)$

3.6 NOTES

The approach in Section 3.1 for proving Theorem 3.9 is inspired by [5] and [28]. In the latter reference a proof of Theorem 3.9 is given for a special case. Results on the general case, as studied in the present chapter, are also given in [27], which is the report version of [28]. See also the lecture notes [37]. Sections 3.2 and 3.3 are inspired by [23], in which the notion of a generalized eigenmode is introduced. Theorem 3.15 is related to the results obtained in [34]. The explicit expressions and proofs in these chapters follow directly from the standard theory but have not appeared elsewhere. The expressions can be combined with well-known methods to obtain a normal form of a square regular matrix, and they can be used with techniques from previous chapters for computing the eigenvalue and eigenvector of each of the diagonal matrices. Then, using the reduced graph, the generalized eigenmode can be obtained as described in the proofs of Theorem 3.15 and Corollary 3.16. In Chapter 6 an algorithmic approach will be presented based on an approach using bouquet matrices.

The notions of sunflower and bouquet matrices are new. They appear for the first time in this book. Though there are flowers of similar shape, we chose the name *sunflower* because of the famous painting by Vincent van Gogh. In botanic terms, the unique circuit of a sunflower graph corresponds to the disc flower, the part of the sunflower with all the seeds. In the same way, the paths leaving from this circuit correspond to the petals (which together form the corolla).

Chapter Four

Asymptotic Qualitative Behavior

As in the previous chapter, we will study in this chapter sequences $\{x(k) : k \in \mathbb{N}\}$ given through

$$x(k+1) = A \otimes x(k), \qquad k \in \mathbb{N}, \tag{4.1}$$

with initial vector $x(0) = x_0 \in \mathbb{R}_{\max}^n$ and $A \in \mathbb{R}_{\max}^{n \times n}$. Provided that A is irreducible with unique eigenvalue λ and associated eigenvector v, it follows for $x(0) = v$ and $k \geq 0$ that $x(k) = A^{\otimes k} \otimes x(0) = \lambda^{\otimes k} \otimes v$. In words, the vectors $x(k)$ are proportional to v, and we may therefore say that the *qualitative* asymptotic behavior of $x(k)$ is completely characterized by v. We have already encountered this type of limit in the heap model as described in Section 1.3. The qualitative limiting behavior of $x(k)$ falls into one of two possible scenarios: $x(k)$ reaches the eigenspace of A and behaves according to $x(k+1) = \lambda \otimes x(k)$ for k sufficiently large, or $x(k)$ enters into a periodic regime (to be defined below). Hence, there are two sources of a possible nonuniqueness of the limiting behavior of $x(k)$:

- $x(k)$ enters the eigenspace and this space is of dimension two or higher (there are several nonproportional eigenvectors), or
- $x(k)$ enters a periodic regime.

This chapter is organized as follows. In Section 4.1 the concept of a periodic regime is introduced, and the close relation between periodic regimes, eigenvalues, and eigenvectors is discussed. The eigenspace of irreducible matrices is studied in Section 4.2. A class of matrices with unique qualitative behavior is discussed in Section 4.3. Section 4.4 deals with a proper limit concept for max-plus sequences by means of the projective space. Eventually, in Section 4.5 it is shown that higher-order max-plus recurrence relations can be reduced to first-order ones.

4.1 PERIODIC REGIMES

Let $A \in \mathbb{R}_{\max}^{n \times n}$. A *periodic regime* is a set of vectors $x^1, \ldots, x^d \in \mathbb{R}_{\max}^n$ for some $d \geq 1$ such that a finite number ρ exists that satisfies

$$\rho \otimes x^1 = A \otimes x^d \quad \text{and} \quad x^{i+1} = A \otimes x^i, \qquad i \in \underline{d-1}.$$

If $\overline{x^i} \neq \overline{x^j}$ for $i, j \in \underline{d-1}$ with $i \neq j$, then $x^1, \ldots, x^d$ is said to be of *period d*. A consequence of the above definition is that $x^1, \ldots, x^d$ are eigenvectors of $A^{\otimes d}$ associated with eigenvalue ρ (see exercise 1). If A is irreducible with cyclicity $\sigma(A)$, then A will possess periodic regimes of period $\sigma(A)$ or less. In this context, one may wonder whether the fact that ρ is an eigenvalue of $A^{\otimes d}$ implies that $(1/d) \times \rho$ is an eigenvalue of A. The following theorem gives a positive answer. Moreover, it shows that an eigenvector can be found via a periodic regime.

THEOREM 4.1 *Let* $x^1, \ldots, x^d$ *be a periodic regime for matrix* A *with* $\rho \otimes x^1 = A \otimes x^d$. *Then* A *has an eigenvalue* λ *that satisfies* $\rho = \lambda^{\otimes d}$, *and a corresponding eigenvector* v *is given by*

$$v = \bigoplus_{j=1}^{d} \lambda^{\otimes(d-j)} \otimes x^j.$$

Proof. We prove the theorem by showing that $A \otimes v = \lambda \otimes v$. Indeed, we have that

$$A \otimes v = A \otimes \left(\bigoplus_{j=1}^{d} \lambda^{\otimes(d-j)} \otimes x^j \right)$$

$$= \bigoplus_{j=1}^{d} A \otimes \lambda^{\otimes(d-j)} \otimes x^j.$$

Noticing that $A \otimes x^j = x^{j+1}$ and $A \otimes x^d = \lambda^{\otimes d} \otimes x^1$ yields

$$\bigoplus_{j=1}^{d} A \otimes \lambda^{\otimes(d-j)} \otimes x^j = \lambda^{\otimes d} \otimes x^1 \oplus \bigoplus_{j=1}^{d-1} \lambda^{\otimes(d-j)} \otimes x^{j+1}$$

$$= \lambda^{\otimes d} \otimes x^1 \oplus \bigoplus_{l=2}^{d} \lambda^{\otimes(d-l+1)} \otimes x^l$$

$$= \bigoplus_{l=1}^{d} \lambda^{\otimes(d-l+1)} \otimes x^l$$

$$= \lambda \otimes \left(\bigoplus_{l=1}^{d} \lambda^{\otimes(d-l)} \otimes x^l \right)$$

$$= \lambda \otimes v,$$

which proves the claim. □

The above theorem is illustrated with the following example.

Example 4.1.1 *Consider the matrix given in (0.10):*

$$A = \begin{pmatrix} 2 & 5 \\ 3 & 3 \end{pmatrix}.$$

Taking $x(0) = (0,0)^\top$ *yields the sequence as given in (0.3):*

$$x(1) = \begin{pmatrix} 5 \\ 3 \end{pmatrix}, \ x(2) = \begin{pmatrix} 8 \\ 8 \end{pmatrix}, \ x(3) = \begin{pmatrix} 13 \\ 11 \end{pmatrix}, \ x(4) = \begin{pmatrix} 16 \\ 16 \end{pmatrix}, \ldots,$$

which is a periodic regime of period 2 with $\rho = 8$. *In particular,*

$$x(k) \in \overline{\begin{pmatrix} 2 \\ 0 \end{pmatrix}}, \qquad k \in \{1, 3, 5, \ldots\},$$

and

$$x(k) \in \overline{\begin{pmatrix} 0 \\ 0 \end{pmatrix}}, \qquad k \in \{0, 2, 4, \ldots\}.$$

By Theorem 4.1, $\lambda = \rho/2 = 4$, which can be easily checked by inspecting the communication graph of A. Indeed, $\mathcal{G}(A)$ consists of the elementary circuits $(1, 1)$, $(2, 2)$, and $((1, 2), (2, 1))$. The circuit $((1, 2), (2, 1))$ is critical with average weight equal to 4. Theorem 4.1 also yields an eigenvector of A. Take

$$x^1 = \begin{pmatrix} 5 \\ 3 \end{pmatrix}, \quad x^2 = \begin{pmatrix} 8 \\ 8 \end{pmatrix};$$

then

$$\lambda^{\otimes 1} \otimes x^1 \oplus \lambda^{\otimes 0} \otimes x^2 = 4 \otimes \begin{pmatrix} 5 \\ 3 \end{pmatrix} \oplus 0 \otimes \begin{pmatrix} 8 \\ 8 \end{pmatrix} = \begin{pmatrix} 9 \\ 8 \end{pmatrix}$$

yields an eigenvector of A.

4.2 CHARACTERIZATION OF THE EIGENSPACE

Let A have finite eigenvalue λ. The eigenspace $V(A, \lambda)$ of matrix A is the set of all eigenvectors of A corresponding to λ, and $V(A, \lambda)$ obviously is a linear space. The eigenspaces of A and A_λ coincide. Indeed, for $v \in V(A)$, it holds for any j that

$$[\lambda \otimes v]_j = [A \otimes v]_j \Leftrightarrow v_j = [A \otimes v]_j - \lambda \Leftrightarrow e \otimes v_j = [A_\lambda \otimes v]_j. \quad (4.2)$$

In fact, for any β, the eigenspaces of A and A_β coincide. Hence, we may conveniently work with either A or A_λ for the analysis of the eigenspace of A. Recall that we write $[A]_{.k}$ to indicate the kth column of A and that we denote the nodes of the critical graph by $\mathcal{N}^c(A)$, i.e., $\eta \in \mathcal{N}^c(A)$ if η lies on a critical circuit. Given an eigenvalue λ and an associated eigenvector v, the *saturation graph*, denoted by $\mathcal{S}(A, v)$, consists of those arcs (j, i) of $\mathcal{G}(A)$ such that $a_{ij} \otimes v_j = \lambda \otimes v_i$ with $v_i, v_j \neq \varepsilon$. Of course, the saturation graph $\mathcal{S}(A, v)$ also depends on λ in principle, but that will not be reflected in the notation.

LEMMA 4.2 *For $A \in \mathbb{R}_{\max}^{n \times n}$ with finite eigenvalue λ and associated finite eigenvector v, the following hold:*

- *For each node i in $\mathcal{S}(A, v)$, there exists a circuit in $\mathcal{S}(A, v)$ from which node i can be reached in a finite number of steps.*
- *The circuits of any saturation graph have average circuit weight λ.*
- *If, in addition, A is irreducible, then circuits of any saturation graph belong to the critical graph.*

Proof. Let i be a node of the saturation graph. Then there exists a node j in the saturation graph such that $\lambda \otimes v_i = a_{ij} \otimes v_j$ with $v_i, v_j \neq \varepsilon$. Repeating this argument, we find a node k such that $\lambda \otimes v_j = a_{jk} \otimes v_k$ with $v_j, v_k \neq \varepsilon$. Repeating this argument an arbitrary number of times, say, m, we get a path in $\mathcal{S}(A,v)$ of length m. If $m > n$, the constructed path must contain a circuit.

We turn to the proof of the second part of the lemma. Let

$$\rho = ((i_1, i_2), (i_2, i_3), \ldots, (i_l, i_{l+1} = i_1))$$

be a circuit in $\mathcal{S}(A,v)$. By definition,

$$\lambda \otimes v_{i_{k+1}} = a_{i_{k+1}i_k} \otimes v_{i_k},$$

which implies

$$\lambda^{\otimes l} \otimes v_{i_1} = \bigotimes_{k=1}^{l} a_{i_{k+1}i_k} \otimes v_{i_1}.$$

Hence,

$$\lambda^{\otimes l} = \bigotimes_{k=1}^{l} a_{i_{k+1}i_k},$$

but the expression on the right-hand side of the above formula is the weight of the circuit ρ, which thus has average weight λ.

Finally, we deal with the third part of the lemma. Theorem 2.9 implies that λ (the eigenvalue of A) equals the maximal average circuit weight of $\mathcal{G}(A)$. According to the second part of the lemma, circuits in the saturation graph have average circuit weight λ. Hence, the average weight of any circuit in the saturation graph equals the maximal average circuit weight of $\mathcal{G}(A)$, and the circuit thus belongs to the critical graph. □

LEMMA 4.3 *Let $A \in \mathbb{R}_{\max}^{n \times n}$ be irreducible and have eigenvalue λ, and let v be an eigenvector associated with λ. Then, A_λ^* has eigenvalue e, and v is an associated eigenvector.*

Proof. Note that

$$(E \oplus A_\lambda) \otimes v = v \quad \text{and} \quad A_\lambda^* = (E \oplus A_\lambda)^{\otimes(n-1)};$$

see exercise 9. Hence,

$$A_\lambda^* \otimes v = (E \oplus A_\lambda)^{\otimes(n-1)} \otimes v = v.$$

□

As the following lemma shows, for an irreducible matrix A, the vectors $[A_\lambda^*]_{\cdot i}$, with $i \in \mathcal{N}^c(A)$, constitute a basis of the eigenspace of A.

LEMMA 4.4 *Let $A \in \mathbb{R}_{\max}^{n \times n}$ be irreducible with eigenvalue λ, and let v be an eigenvector associated with λ. Then v can be written as*

$$v = \bigoplus_{i \in \mathcal{N}^c(A)} a_i \otimes [A_\lambda^*]_{\cdot i},$$

with $a_i \in \mathbb{R}_{\max}$.

Proof. By Lemma 4.3, A_λ^* has eigenvalue e, and provided that v is an eigenvector of A associated to λ, v is also an eigenvector of A_λ^* for eigenvalue e.

Consider two nodes i, j in $\mathcal{S}(A_\lambda, v)$ such that there exists a path from i to j, say, $((i_1, i_2), (i_2, i_3), \dots, (i_l, i_{l+1}))$, with $i = i_1$ and $j = i_{l+1}$. This gives

$$[A_\lambda]_{i_{k+1} i_k} \otimes v_{i_k} = v_{i_{k+1}}, \qquad k \in \underline{l}.$$

Hence, $v_j = a \otimes v_i$ with

$$a = \bigotimes_{k=1}^{l} [A_\lambda]_{i_{k+1} i_k} \le \left[A_\lambda^{\otimes l}\right]_{ji} \le [A_\lambda^*]_{ji}. \qquad (4.3)$$

Using the fact that $v_j = a \otimes v_i$ yields for an arbitrary node $\eta \in \underline{n}$

$$
\begin{aligned}
[A_\lambda^*]_{\eta j} \otimes v_j &= [A_\lambda^*]_{\eta j} \otimes a \otimes v_i \\
&\overset{(4.3)}{\le} [A_\lambda^*]_{\eta j} \otimes [A_\lambda^*]_{ji} \otimes v_i \\
&\le [A_\lambda^*]_{\eta i} \otimes v_i, \qquad (4.4)
\end{aligned}
$$

where the last inequality follows from $A_\lambda^* \otimes A_\lambda^* = A_\lambda^*$ (see exercise 3). By applying Lemma 4.2 (with the roles of i and j interchanged) for any j in the saturation graph, there exists a node $i = i(j)$ that belongs to a critical circuit. Inequality (4.4) therefore implies

$$\bigoplus_{j \in \mathcal{S}(A_\lambda, v)} [A_\lambda^*]_{\eta j} \otimes v_j \le \bigoplus_{i \in \mathcal{N}^c(A_\lambda)} [A_\lambda^*]_{\eta i} \otimes v_i, \qquad (4.5)$$

for $\eta \in \underline{n}$.

The basic equation for the eigenvector v is $v = A_\lambda^* \otimes v$. By definition, the value of v_η is equal to $[A_\lambda^*]_{\eta j} \otimes v_j$ for some j that has to be in the saturation graph. We do not know which specific node j determines v_η, but since the node has to be in the saturation graph, it holds for $\eta \in \underline{n}$ that

$$v_\eta = \bigoplus_{j \in \mathcal{S}(A_\lambda, v)} [A_\lambda^*]_{\eta j} \otimes v_j \overset{(4.5)}{\le} \bigoplus_{j \in \mathcal{N}^c(A_\lambda)} [A_\lambda^*]_{\eta j} \otimes v_j.$$

Conversely, since v is an eigenvector of A_λ^* for eigenvalue e,

$$v_\eta = [A_\lambda^* \otimes v]_\eta = \bigoplus_{j=1}^{n} [A_\lambda^*]_{\eta j} \otimes v_j \ge \bigoplus_{i \in \mathcal{N}^c(A_\lambda)} [A_\lambda^*]_{\eta i} \otimes v_i$$

for $\eta \in \underline{n}$, which completes the proof of the lemma. Specifically, we obtain $a_i = v_i$ for $i \in \mathcal{N}^c(A_\lambda)$. In case some of the columns of A_λ^* are colinear, then the a_i's are nonunique and some can be chosen equal to ε. Noticing that $\mathcal{N}^c(A_\lambda) = \mathcal{N}^c(A)$ concludes the proof. $\qquad \square$

The following theorem characterizes the eigenspace of an irreducible square matrix.

THEOREM 4.5 *Let A be irreducible, and let A_λ^* be defined as in (2.10).*

(i) *If node i belongs to the critical graph, then $[A_\lambda^*]_{\cdot i}$ is an eigenvector of A.*

(ii) *The eigenspace of A is given by*

$$V(A) = \left\{ v \in \mathbb{R}^n_{\max} : v = \bigoplus_{i \in \mathcal{N}^c(A)} a_i \otimes [A^*_\lambda]_{\cdot i} \quad \text{for } a_i \in \mathbb{R}_{\max} \right\}.$$

(iii) *For i, j belonging to the critical graph, there exists $a \in \mathbb{R}$ such that*

$$a \otimes [A^*_\lambda]_{\cdot i} = [A^*_\lambda]_{\cdot j} \tag{4.6}$$

if and only if i and j belong to the same m.s.c.s. of the critical graph.

Proof. Note that irreducibility of A implies regularity of A and therefore part (i) has already been proved in Lemma 2.7.

For the proof of part (ii), notice that the columns $[A^*_\lambda]_{\cdot i}$ for $i \in \mathcal{N}^c(A)$ are eigenvectors of A, which is part (i) of the theorem. Since any linear combination of eigenvectors is again an eigenvector, it follows that

$$\bigoplus_{i \in \mathcal{N}^c(A)} a_i \otimes [A^*_\lambda]_{\cdot i},$$

with $a_i \in \mathbb{R}_{\max}$ and at least one a_i finite, is an eigenvector of A. Conversely, any eigenvector can be written as a linear combination of the columns $[A^*_\lambda]_{\cdot i}$ for $i \in \mathcal{N}^c(A)$, which has been shown in Lemma 4.4. This proves part (ii).

We now turn to the proof of part (iii). We first prove that if i and j belong to the same m.s.c.s., then (4.6) holds. Subsequently it will be shown that if i and j do not belong to the same m.s.c.s., then (4.6) cannot hold.

If i and j belong to the same m.s.c.s. of the critical graph of A_λ, then $[A^*_\lambda]_{ji} \otimes [A^*_\lambda]_{ij} = e$ and, hence,

$$[A^*_\lambda]_{li} \otimes [A^*_\lambda]_{ij} \leq [A^*_\lambda]_{lj}$$
$$= [A^*\lambda]_{lj} \otimes [A^*_\lambda]_{ji} \otimes [A^*_\lambda]_{ij}$$
$$\leq [A^*_\lambda]_{li} \otimes [A^*_\lambda]_{ij}, \qquad \forall l \in \underline{n},$$

which shows that

$$[A^*_\lambda]_{li} \otimes [A^*_\lambda]_{ij} = [A^*_\lambda]_{lj}, \qquad \forall l \in \underline{n}.$$

Hence, (4.6) has been proved with $a = [A^*_\lambda]_{ij}$.

Now suppose that i and j do not belong to the same m.s.c.s., but that notwithstanding (4.6) holds true. The ith and jth components of (4.6) read

$$a \otimes e = [A^*_\lambda]_{ij} \quad \text{and} \quad a \otimes [A^*_\lambda]_{ji} = e,$$

respectively, from where it follows that

$$[A^*_\lambda]_{ij} \otimes [A^*_\lambda]_{ji} = e.$$

As a consequence, the circuit formed by the arcs (i, j) and (j, i) has average weight e, and therefore, nodes i and j belong to the critical graph of A and, hence, belong to the same m.s.c.s. (of this critical graph). Thus, a contradiction has been obtained. $\square$

The application of the above theorem is illustrated with the following series of examples.

Example 4.2.1   *Consider the matrix*

$$A = \begin{pmatrix} 2 & 5 \\ 3 & 3 \end{pmatrix}.$$

Matrix A is irreducible with unique eigenvalue $\lambda = \lambda(A) = 4$, and the critical graph of A consists of the nodes $\{1,2\}$. The critical graph has thus one m.s.c.s. and $\sigma(A) = 2$. By computation, it follows that

$$A_\lambda^* = \begin{pmatrix} 0 & 1 \\ -1 & 0 \end{pmatrix}.$$

By Theorem 4.5, the vectors

$$\begin{pmatrix} 0 \\ -1 \end{pmatrix} \quad \text{and} \quad \begin{pmatrix} 1 \\ 0 \end{pmatrix}$$

are eigenvectors of A. Clearly, these vectors are max-plus multiples of each other. This also follows from Theorem 4.5 since the associated nodes/indices belong to the same m.s.c.s. of the critical graph. Hence, it follows that the eigenspace of A is given by

$$V(A) = \left\{ v \in \mathbb{R}_{\text{max}}^2 : v = a \otimes \begin{pmatrix} 1 \\ 0 \end{pmatrix} \text{ for } a \in \mathbb{R} \right\}$$

or, more concisely,

$$V(A) = \overline{\begin{pmatrix} 1 \\ 0 \end{pmatrix}}.$$

Compare the above result with Example 4.1.1. There, we computed an eigenvector of A but had no information whether the obtained eigenvector already character-ized the complete eigenspace of A. Theorem 4.5 shows that the eigenspace of A is of dimension one. Moreover, it provides an algebraic way of computing all vectors in $V(A)$.

Example 4.2.2   *Consider the matrix*

$$A = \begin{pmatrix} 0 & -2 \\ 1 & 0 \end{pmatrix}.$$

Matrix A is irreducible with a unique eigenvalue $\lambda = \lambda(A) = 0$, and the critical graph of A consists of the nodes $\{1,2\}$ and circuits (or self-loops) $(1,1)$ and $(2,2)$. Thus, the critical graph has two m.s.c.s.'s, namely, the circuits $(1,1)$ and $(2,2)$, and $\sigma(A) = 1$. For A it holds that

$$A = A^{\otimes k} = A_\lambda = A^+ = A^* = A_\lambda^*, \qquad k \in \mathbb{N}.$$

By Theorem 4.5, the vectors

$$\begin{pmatrix} 0 \\ 1 \end{pmatrix} \quad \text{and} \quad \begin{pmatrix} -2 \\ 0 \end{pmatrix}$$

are eigenvectors of A. Obviously, these vectors are not max-plus multiples of each other. This fact also follows from Theorem 4.5 since the associated nodes/indices belong to different m.s.c.s.'s of the critical graph. By Theorem 4.5 it further follows that the eigenspace of A is given by

$$V(A) = \left\{ v \in \mathbb{R}^2_{\max} : v = a_1 \otimes \begin{pmatrix} 0 \\ 1 \end{pmatrix} \oplus a_2 \otimes \begin{pmatrix} -2 \\ 0 \end{pmatrix} \text{ for } a_1, a_2 \in \mathbb{R} \right\}.$$

Example 4.2.3 *Consider A in Example 3.1.3. The critical graph of A has one m.s.c.s. consisting of the circuit $((1,2),(2,1))$. It follows that $\lambda = \lambda(A) = 6$. Application of Theorem 4.5 yields*

$$A^*_\lambda = \begin{pmatrix} 0 & 5 \\ -5 & 0 \end{pmatrix} \quad \text{and} \quad V(A) = \overline{\begin{pmatrix} 5 \\ 0 \end{pmatrix}}.$$

4.3 PRIMITIVE MATRICES

From a system-theoretical point of view, we are interested in the limiting behavior of $x(k)$. More precisely, we are interested in the behavior of $\overline{x(k)}$ for k large. Let A in (4.1) be irreducible. Theorem 3.9 and the text thereafter imply that for any initial vector x_0 it holds that $x(k)$ enters after at most $t(A)$ iterations a periodic regime with period $\sigma(A)$ or smaller. If A has cyclicity one, then $x(k)$ enters after at most $t(A)$ iterations the eigenspace of A; in formula, we write

$$x(k) \in V(A) \quad \text{for } k \geq t(A).$$

Generally speaking, there are two possible scenarios: $x(k)$ enters the eigenspace, or $x(k)$ enters an orbit whose period is larger than one and maximally $\sigma(A)$. We call the set of all initial conditions x_0 such that $A^{\otimes k} \otimes x_0$ eventually reaches the set $\overline{v}$ (resp., the periodic regime $x^1, \ldots, x^d$) for some eigenvector v the *domain of attraction* of v (resp., $x^1, \ldots, x^d$), with $d > 1$. For example, for the matrix given in Example 4.1.1, the vector $x = (0,0)^\top$ lies in the domain of attraction of the periodic regime $(5,3)^\top, (8,8)^\top$.

Let A be irreducible, and suppose that the critical graph of A has a single m.s.c.s. of cyclicity one. Such matrices are called *primitive*. Primitive matrices form an important class of square matrices because for any primitive matrix A the sequence $\{A^{\otimes k} \otimes x_0 : k \in \mathbb{N}\}$ has, independent of the initial vector $x_0 \in \mathbb{R}^n$, a unique limit behavior. A way of expressing this is by saying that $V(A) = \overline{v}$ for some $v \in \mathbb{R}^n$ and that the domain of attraction of v is equal to $\mathbb{R}^n$. More explicitly, let A be primitive and let $x(k) = A^{\otimes k} \otimes x_0$ for $k \geq 0$, with $x_0 \in \mathbb{R}^n$; then, independently of the initial vector x_0, the following hold:

- The asymptotic growth rate is equal to the eigenvalue of A:

$$\lim_{k \to \infty} \frac{x_j(k)}{k} = \lambda, \qquad j \in \underline{n},$$

where λ is the eigenvalue of A (see Lemma 3.12).

- There exists a unique $\overline{v} \in \mathbb{PR}^n$ such that

$$\forall k \geq t(A) : \qquad \overline{v} = \overline{x(k)}$$

(see exercise 2).

A way of phrasing the latter point is by saying that $x(k)$ reaches after at most $t(A)$ iterations its asymptotic regime; any influence of the initial vector has died out and the growth rate of $x(k)$ coincides with the eigenvalue of A.

Example 4.3.1 *Let $A \in \mathbb{R}_{\max}^{n \times n}$ model a railway network with n stations. Suppose that A is irreducible with eigenvalue λ. Then any eigenvector v of A provides a regular timetable; that is, initializing x_0 to v, the kth departure of a train from station j happens at time $\lambda^{\otimes k} \otimes v_j = k \times \lambda + v_j$. If A is irreducible and its critical graph has more than one m.s.c.s., several nonproportional regular timetables exist; that is, $v, w \in V(A)$ exist such that $\overline{v} \neq \overline{w}$, and initializing x_0 to either v or w again results in a regular behavior for the departure time of the kth train from station j, namely, $\lambda^{\otimes k} \otimes v_j = k \times \lambda + v_j$ or $\lambda^{\otimes k} \otimes w_j = k \times \lambda + w_j$, respectively.*

Suppose that we intend to operate the network with timetable $v \in V(A)$, but unfortunately—due to some external influence—the first trains can only start at time vector $z \geq v$ (with at least one component being a strict inequality). If A is of cyclicity 2 or higher, then $x(k + 1) = A \otimes x(k)$, with $x_0 = z$, may enter a periodic regime of period $\sigma(A)$. Formally, z may lie in the domain of attraction of a periodic regime with period d, where $1 < d \leq \sigma(A)$. In words, the network will not completely recover from the initial delay, and no regular timetable is reached. If, however, A is irreducible and primitive, then there is a unique timetable (up to scalar multiplication), and any initial delay will eventually die out and the network will return to its unique regular timetable (apart from possibly the same constant shift in all departure times). This kind of phenomenon will be discussed in detail in Section 9.1.

4.4 LIMITS IN THE PROJECTIVE SPACE

On $\mathbb{R}^n$, we define

$$||x||_{\mathbb{P}} \stackrel{\text{def}}{=} \bigoplus_{i=1}^{n} x_i \otimes \bigoplus_{i=1}^{n} (-x_i) = \max_{i \in \underline{n}} x_i - \min_{i \in \underline{n}} x_i.$$

It is easy to check that $||x||_{\mathbb{P}}$ is the same for all vectors that belong to $\overline{x}$ (see exercise 4), and we define $||\overline{x}||_{\mathbb{P}} = ||x||_{\mathbb{P}}$. Then $||\overline{x}||_{\mathbb{P}} \geq 0$ for any $\overline{x} \in \mathbb{PR}^n$, and

$$||\overline{x}||_{\mathbb{P}} = 0 \quad \text{if and only if} \quad \overline{x} = \overline{\mathbf{u}};,$$

that is, $||\overline{x}||_{\mathbb{P}} = 0$ if and only if for any $x \in \overline{x}$ it holds that all components are equal. For $\alpha \in \mathbb{R}$, let $\alpha \times x$ be defined as the componentwise conventional multiplication of x by α. Thus, $\alpha \times \overline{x} = \overline{\alpha \times x}$, which implies

$$||\alpha \times \overline{x}||_{\mathbb{P}} = |\alpha| \times ||\overline{x}||_{\mathbb{P}}$$

for $\alpha \in \mathbb{R}$ and $\overline{x} \in \mathbb{PR}^n$, where $|\alpha|$ denotes the absolute value of λ. Expression $||\cdot||_{\mathbb{P}}$ also satisfies the triangular inequality. To see this, let $\overline{x}, \overline{y} \in \mathbb{PR}^n$; then, for any $x \in \overline{x}$ and $y \in \overline{y}$,

$$
\begin{aligned}
||\overline{x} + \overline{y}||_{\mathbb{P}} &= \max_i(x_i + y_i) - \min_j(x_j + y_j) \\
&\leq \max_i(\max_k(x_k) + y_i) - \min_j(\min_l(x_l) + y_j) \\
&= \max_k x_k - \min_l x_l + \max_i y_i - \min_j y_j \\
&= ||\overline{x}||_{\mathbb{P}} + ||\overline{y}||_{\mathbb{P}}.
\end{aligned}
$$

Hence, $||\cdot||_{\mathbb{P}}$ is a norm on $\mathbb{PR}^n$. Using the convention in (2.7), we extend the definition of $||\cdot||_{\mathbb{P}}$ to $\mathbb{PR}^n_{\max}$. However, according to the strict use in conventional algebra, $||\cdot||_{\mathbb{P}}$ fails to be a norm on $\mathbb{PR}^n_{\max}$. For any $x \in \mathbb{R}^n_{\max}$ with at least one finite element and at least one element equal to ε, it holds that $||\overline{x}||_{\mathbb{P}} = \infty$, whereas a norm is by definition a mapping to $\mathbb{R}$.

On $\mathbb{PR}^n$, let $x - y$ be defined as the componentwise conventional difference of x and y. With this definition, we obtain a metric $d_{\mathbb{P}}(\cdot, \cdot)$ on $\mathbb{PR}^n$ in the natural way. For $x \in \overline{x}$ and $y \in \overline{y}$, set

$$
d_{\mathbb{P}}(\overline{x}, \overline{y}) = ||\overline{x} - \overline{y}||_{\mathbb{P}}
$$

or, more explicitly,

$$
\begin{aligned}
d_{\mathbb{P}}(\overline{x}, \overline{y}) &= \bigoplus_{i=1}^{n}(x_i - y_i) \otimes \bigoplus_{j=1}^{n}(y_j - x_j) \\
&= \max_{i \in \underline{n}}(x_i - y_i) - \min_{j \in \underline{n}}(x_j - y_j) \\
&= \max_{j \in \underline{n}}(y_j - x_j) - \min_{i \in \underline{n}}(y_i - x_i),
\end{aligned}
$$

where for the last equality we have used that $\max_i(x_i - y_i) = -\min_i(y_i - x_i)$. The metric $d_{\mathbb{P}}(\cdot, \cdot)$ is called the *projective metric*. We extend the definition of $d_{\mathbb{P}}(\cdot, \cdot)$ to $\mathbb{PR}^n_{\max}$ by adopting the convention that $\varepsilon - x = \varepsilon$ for $x \neq \varepsilon$. Note that $d_{\mathbb{P}}(\cdot, \cdot)$ fails to be a metric on $\mathbb{PR}^n_{\max}$. To see this, let $\overline{y}$ be such that all components of y are equal to ε. Then, for any $\overline{x} \in \mathbb{PR}^n$, it follows that $d_{\mathbb{P}}(\overline{x}, \overline{y}) = 0$, whereas $\overline{x} \neq \overline{y}$.

We can now give a precise definition for the limit in (1.14) in Section 1.4. We say that $\overline{x(k)}$ converges to $\overline{x}$ as k tends to ∞ if for any $\delta > 0$ a number $K \in \mathbb{N}$ exists such that for all $k \geq K$ it holds that

$$
d_{\mathbb{P}}(\overline{x}(k), \overline{x}) < \delta.
$$

Example 4.4.1   *Consider the sequence*

$$
x(k) = \begin{pmatrix} k + \frac{k}{2+k} \\ k \end{pmatrix}, \qquad k \geq 1.
$$

Then, in conventional analysis

$$
\lim_{k \to \infty} x(k) = \begin{pmatrix} \infty \\ \infty \end{pmatrix}.
$$

Notice that the above limit fails to capture the fact that $x_1(k) - x_2(k)$ tends to 1 as k tends to ∞.

Consider now the above limit in the projective space; that is, study the (projective) limit of the sequence

$$\overline{x(k)} = \overline{\begin{pmatrix} \frac{k}{2+k} \\ 0 \end{pmatrix}}, \qquad k \in \mathbb{N},$$

as k tends to ∞. Since

$$\overline{\begin{pmatrix} k+1 \\ k \end{pmatrix}} = \overline{\begin{pmatrix} 1 \\ 0 \end{pmatrix}}, \qquad k \in \mathbb{N},$$

it follows that

$$d_{\mathbb{P}}\left(\overline{x(k)}, \overline{\begin{pmatrix} 1 \\ 0 \end{pmatrix}}\right) = d_{\mathbb{P}}\left(\overline{\begin{pmatrix} k + \frac{k}{2+k} \\ k \end{pmatrix}}, \overline{\begin{pmatrix} k+1 \\ k \end{pmatrix}}\right)$$

$$= \max\left(\frac{-2}{2+k}, 0\right) - \min\left(\frac{-2}{2+k}, 0\right)$$

$$= \frac{2}{2+k},$$

which becomes arbitrarily small. Hence, $\overline{x(k)}$ converges to $\overline{(1,0)^{\mathsf{T}}}$ as k tends to ∞. The limit in the projective space keeps track of the finite differences within the vector $x(k)$. Even though $x(k)$ diverges (in the conventional sense), convergence of $x(k)$ in the projective space to a finite element happens.

4.5 HIGHER-ORDER RECURRENCE RELATIONS

For $M \geq 0$, let $A_m \in \mathbb{R}_{\max}^{n \times n}$ for $0 \leq m \leq M$ and $x(m) \in \mathbb{R}_{\max}^n$ for $-M \leq m \leq -1$. Then, the (implicit) recurrence relation

$$x(k) = \bigoplus_{m=0}^{M} A_m \otimes x(k-m), \qquad k \geq 0, \tag{4.7}$$

is defined. The above recurrence relation is called an Mth-order recurrence relation. So far we have restricted our analysis to first-order recurrence relations with $A_0 = \mathcal{E}$. However, in applications one frequently encounters systems whose dynamics follow a recurrence relation of order two or higher and/or for which $A_0 \neq \mathcal{E}$. As we will show in this section, the Mth-order recurrence relation (4.7) can be transformed into a first-order recurrence relation of the type

$$x(k+1) = A \otimes x(k), \qquad k \geq 0,$$

provided that A_0 in (4.7) has circuit weights less than or equal to zero or has no circuits at all. The starting point is Lemma 2.2, which implies that if A_0 has circuit weights less than or equal to zero, then

$$A_0^* = \bigoplus_{i=0}^{n-1} A_0^{\otimes i}. \tag{4.8}$$

We now turn to the algebraic manipulation of (4.7). Set

$$b(k) = \bigoplus_{m=1}^{M} A_m \otimes x(k - m).$$

Then (4.7) reduces to

$$x(k) = A_0 \otimes x(k) \oplus b(k). \tag{4.9}$$

By Theorem 2.10, (4.9) can be written as

$$x(k) = A_0^* \otimes b(k)$$

or, more explicitly,

$$x(k) = A_0^* \otimes A_1 \otimes x(k - 1) \oplus \cdots \oplus A_0^* \otimes A_M \otimes x(k - M). \tag{4.10}$$

The difference between (4.7) and (4.10) is that in the latter $x(k)$ occurs only in the left-hand side of the equation.

As a next step, we transform (4.10) into a first-order recurrence relation. In order to do so, we set

$$\tilde{x}(k) = (x^\top(k - 1), x^\top(k - 2), \ldots, x^\top(k - M))^\top$$

and

$$\tilde{A} = \begin{pmatrix} A_0^* \otimes A_1 & A_0^* \otimes A_2 & \cdots & \cdots & A_0^* \otimes A_M \\ E & \mathcal{E} & \cdots & \cdots & \mathcal{E} \\ \mathcal{E} & E & \ddots & & \mathcal{E} \\ \vdots & & \ddots & & \vdots \\ \mathcal{E} & \mathcal{E} & \cdots & E & \mathcal{E} \end{pmatrix}.$$

Then, (4.7) can be written as

$$\tilde{x}(k + 1) = \tilde{A}(k) \otimes \tilde{x}(k), \qquad k \geq 0. \tag{4.11}$$

4.6 EXERCISES

1. Let $x^1, \ldots, x^d \in \mathbb{R}^n$ be a periodic regime of period d of $A \in \mathbb{R}_{\max}^{n \times n}$ such that $\mu \otimes x^1 = A \otimes x^d$. Show that $x^1, \ldots, x^d$ are eigenvectors of $A^{\otimes d}$ associated with eigenvalue μ.

2. For $A \in \mathbb{R}_{\max}^{n \times n}$ we say that the *eigenvector of A is unique* if for any two eigenvectors v, w it holds that $v = \alpha \otimes w$ for $\alpha \in \mathbb{R}$. Show that if $A \in \mathbb{R}_{\max}^{n \times n}$ is irreducible and if the critical graph of A consists of a single strongly connected subgraph, then the eigenvector of A is unique.

3. Show that $A_\lambda^* \otimes A_\lambda^* = A_\lambda^*$.

4. Show that $||\overline{x}||_\mathbb{P}$ is independent of the representative x; that is, for $y \in \overline{x}$ it holds that $||\overline{y}||_\mathbb{P} = ||\overline{x}||_\mathbb{P}$.

5. Show that $d_\mathbb{P}(\cdot, \cdot)$ is a metric on $\mathbb{PR}^n$; that is, show that

   - $d_\mathbb{P}(\overline{x}, \overline{y}) = d_\mathbb{P}(\overline{y}, \overline{x})$ for $\overline{x}, \overline{y} \in \mathbb{PR}^n$;

- $d_{\mathbb{P}}(\overline{x}, \overline{y}) = 0$ if and only if $\overline{x} = \overline{y}$;
- $d_{\mathbb{P}}(\overline{x}, \overline{y}) + d_{\mathbb{P}}(\overline{y}, \overline{z}) \geq d_{\mathbb{P}}(\overline{x}, \overline{z})$ for $\overline{x}, \overline{y}, \overline{z} \in \mathbb{PR}^n$.

6. Show that max-plus algebra is nonexpansive in the $d_{\mathbb{P}}$-norm; that is, show that for any regular $A \in \mathbb{R}_{\max}^{n \times n}$ and any $x, y \in \mathbb{R}^n$ it holds that

$$d_{\mathbb{P}}(A \otimes x, A \otimes y) \leq d_{\mathbb{P}}(x, y).$$

7. Let

$$A = \begin{pmatrix} 2 & -2 \\ 1 & 2 \end{pmatrix}.$$

Compute a basis of $V(A)$.

8. Let

$$x(k) = \begin{pmatrix} \varepsilon & 1 \\ -4 & \varepsilon \end{pmatrix} \otimes x(k) \oplus \begin{pmatrix} 2 & 4 \\ -1 & \varepsilon \end{pmatrix} \otimes x(k-2), \qquad k \geq 2,$$

with given $x(1), x(0) \in \mathbb{R}^2$. Transfer the above recurrence relation into a recurrence relation of the form $\tilde{x}(k+1) = \tilde{A} \otimes \tilde{x}(k)$.

9. Let A be irreducible. Show that $A_\lambda^* = (E \oplus A_\lambda)^{\otimes(n-1)}$, where $\lambda = \lambda(A)$ and n denotes the number of rows and columns of A.

10. If λ is an eigenvalue of matrix A with some appropriate eigenvector, then $\lambda^{\otimes k}$ is an eigenvalue of $A^{\otimes k}$ for the same eigenvector. Conversely, if $\mu = \lambda^{\otimes k}$ is an eigenvalue of $A^{\otimes k}$ with eigenvector w, then

$$v = \bigoplus_{i=1}^{k} A^{\otimes(k-i)} \otimes w \otimes \lambda^{\otimes i}$$

is an eigenvector of A corresponding to eigenvalue λ. Prove this.

4.7 NOTES

Primitive matrices are called *scs1-cyc1 matrices* in the literature. The somewhat awkward expression *scs1-cyc1* stems from the general terminology of calling a matrix whose communication graph has m m.s.c.s.'s and that is of cyclity k a *scs-m-cyc-k matrix*. We refer to [61] for a variety of examples of scs-m-cyc-k matrices with $m, k > 1$.

Mairesse provides a nice graphical representation of the domain of attraction in the projective space for dimension three; see [62] and the extended version [61]. In particular, the eigenvector (resp., periodic regime) in whose domain of attraction an initial value x_0 lies can be deduced from a graphical representation of the eigenspace of A in the projective space.

The set $\mathbb{R}_{\max}$ can be equipped with a metric through an exponential lifting. Set $\exp(-\infty) = \exp(\varepsilon) = 0$, then $|x - y|_{\exp} \stackrel{\text{def}}{=} |\exp(x) - \exp(y)|$ yields a metric on $\mathbb{R}_{\max}$, and convergence in $\mathbb{R}_{\max}$ can be defined through $|\cdot|_{\exp}$.

Making use of the nonexpansiveness of max-algebra in the $d_{\mathbb{P}}$-norm, one can give an alternative proof of Lemma 3.12 via arguments in the projective space.

The statement of part (iii) of Theorem 4.5 can be generalized: $[A_\lambda^*]_{\cdot i}$ cannot be expressed as a linear combination of other $[A_\lambda^*]_{\cdot j}$'s with the j's belonging to other m.s.c.s.'s than the one to which i belongs. The proof, beyond the scope of this book, can be found in [5].

Chapter Five

Numerical Procedures for Eigenvalues of Irreducible Matrices

In this chapter we discuss two numerical procedures for irreducible matrices over max-plus algebra. The first one, called *Karp's algorithm*, will be presented in Section 5.1 and yields the eigenvalue of an irreducible matrix. The second one, called a *power algorithm*, to be presented in Section 5.2, yields the eigenvalue and a corresponding eigenvector. Notice that we have already encountered an algorithm for computing the eigenvalue in Chapter 2. Indeed, by Theorem 2.9 the eigenvalue of an irreducible matrix A is equal to the maximal average circuit weight of the communication graph of A.

We start in this chapter from a matrix $A \in \mathbb{R}_{\max}^{n \times n}$ and consider the recurrence relation that plays a central role in this book,

$$x(k+1) = A \otimes x(k), \qquad (5.1)$$

for all $k \geq 0$. If $\lambda \in \mathbb{R}_{\max}$ and $v \in \mathbb{R}_{\max}$ are an eigenvalue and an eigenvector of A, respectively, then the solution of (5.1) for $x(0) = v$ is given by $x(k) = \lambda^{\otimes k} \otimes v$. It has been shown in Section 2.2 that if matrix A is irreducible, both λ and v exist and are finite (see Lemma 2.8 and Theorem 2.9). Moreover, λ is unique.

5.1 KARP'S ALGORITHM

In this section, we present Karp's algorithm for computing the eigenvalue of an irreducible square matrix. The following theorem, which is due to Karp [55], provides a characterization of the eigenvalue that is different from the one in Theorem 2.9.

THEOREM 5.1 *Let* $A \in \mathbb{R}_{\max}^{n \times n}$ *be irreducible with eigenvalue* λ. *Then*

$$\lambda = \max_{i=1,\ldots,n} \min_{k=0,\ldots,n-1} \frac{[A^{\otimes n}]_{ij} - [A^{\otimes k}]_{ij}}{n-k}, \qquad (5.2)$$

where $j \in \underline{n}$ *can be chosen arbitrarily and division has to be understood in conventional algebra.*

Proof. Recall from Theorem 2.9 that λ can be interpreted as the maximal average circuit weight. To prove the current theorem, consider first the case where $\lambda = 0$. Then $\mathcal{G}(A)$ contains no circuits with positive weight, and there is at least one circuit with weight zero. Since $\mathcal{G}(A)$ contains no circuits of positive weight, it follows from Lemma 2.2 that

$$A^* = \bigoplus_{k=0}^{n-1} A^{\otimes k},$$

which reads in conventional algebra as $[A^*]_{ij} = \max_{0 \le k \le n-1} [A^{\otimes k}]_{ij}$, for all $i, j \in \underline{n}$. This implies

$$\min_{0 \le k \le n-1} \left([A^{\otimes n}]_{ij} - [A^{\otimes k}]_{ij} \right) = [A^{\otimes n}]_{ij} - [A^*]_{ij} \le 0,$$

for all $i, j \in \underline{n}$, which gives

$$\min_{0 \le k \le n-1} \frac{[A^{\otimes n}]_{ij} - [A^{\otimes k}]_{ij}}{n-k} = \frac{[A^{\otimes n}]_{ij} - [A^*]_{ij}}{n-k} \le 0,$$

for all $i, j \in \underline{n}$. Now pick any $j \in \underline{n}$, and to complete the proof for $\lambda = 0$, show that an $i \in \underline{n}$ exists such that

$$[A^{\otimes n}]_{ij} - [A^*]_{ij} = 0.$$

To that end, consider a critical circuit, say, ζ, and let l be a node on ζ. Next, consider a maximal weight path from j to l, say, ξ. To simplify the notation, set $\gamma = |\zeta|_l$ and $\tau = |\xi|_l$. It then holds that $[A^{\otimes \tau}]_{lj} = [A^*]_{lj}$, where it is assumed that $0 \le \tau \le n-1$. Extending ξ by m copies of ζ gives a path from j to l of length $\tau + m \times \gamma$. Denote this path by ξ'. Since the circuit ζ has weight zero, it follows by contradiction that $[A^{\otimes(\tau + m \times \gamma)}]_{lj} = [A^*]_{lj}$ for all integers $m \ge 0$. Now take m such that $\tau + m \times \gamma \ge n$. Then ξ' consists of $\tau + m \times \gamma$ nodes. Let the nth node of ξ' be denoted t. Denote ξ_1 for the subpath of ξ' from node j to node t and ξ_2 for the subpath of ξ' from node t to node l. Clearly, the lengths of ξ_1 and ξ_2 are n and $\tau + m \times \gamma - n$, respectively. Since ξ_1 and ξ_2 are part of a path with maximal weight, they themselves are paths of maximal weight. The weights of ξ_1 and ξ_2 are $[A^{\otimes n}]_{tj}$ and $[A^{\otimes(\tau + m \times \gamma - n)}]_{lt}$, respectively, so that

$$[A^*]_{lj} = [A^{\otimes n}]_{tj} + [A^{\otimes(\tau + m \times \gamma - n)}]_{lt}.$$

For the weight of ξ_1 it holds that $[A^{\otimes n}]_{tj} \le [A^*]_{tj}$. To prove the equality, assume that $[A^{\otimes n}]_{tj} < [A^*]_{tj}$, and consider a path, say, ξ_0, from j to t of maximal weight $[A^*]_{tj}$. Next, extend ξ_0 by the subpath ξ_2 from t to l. The weight of the path from j to l thus obtained is

$$[A^*]_{tj} + [A^{\otimes(\tau + m \times \gamma - n)}]_{lt},$$

which will be larger than $[A^*]_{lj}$ and which by definition is impossible. Hence, $[A^{\otimes n}]_{ij} = [A^*]_{ij}$ with $i = t$, and (5.2) is correct in the case where the maximal circuit mean is zero.

To conclude the proof, assume that λ is finite but not necessarily zero. Subtracting a constant c from each of the entries of A, it follows that λ, seen as a critical circuit mean, is also reduced by c. Further, $[A^{\otimes k}]_{ij}$ is reduced by $k \times c$, implying that

$$\frac{1}{n-k} \left([A^{\otimes n}]_{ij} - [A^{\otimes k}]_{ij} \right)$$

is reduced by c, for all $i, j \in \underline{n}$ and $k \geq 0$. This means that for any $j \in \underline{n}$

$$\min_{0 \leq k \leq n-1} \frac{[A^{\otimes n}]_{ij} - [A^{\otimes k}]_{ij}}{n - k}$$

is reduced by c. Hence, both sides in (5.2) are reduced by c. Taking c equal to λ, the above case, where the maximum circuit mean was supposed to be zero, can be applied.   □

An efficient way of evaluating the eigenvalue of an irreducible matrix with the help of Karp's theorem is the following. Take $x(0) = e_j$ (the jth base vector), and determine $x(k)$ by iterating (5.1). Then $x(k)$ is equal to $[A^{\otimes k}]_{.j}$, for $k \geq 0$. Karp's algorithm can now be stated as follows.

Algorithm 5.1.1 KARP'S ALGORITHM

1. *Choose arbitrary $j \in \underline{n}$, and set $x(0) = e_j$.*

2. *Compute $x(k)$ for $k = 0, \ldots, n$.*

3. *Compute as an eigenvalue*

$$\lambda = \max_{i=1,\ldots,n} \min_{k=0,\ldots,n-1} \frac{x_i(n) - x_i(k)}{n - k}.$$

Karp's algorithm is illustrated with the following examples.

Example 5.1.1   *Let*

$$A = \begin{pmatrix} \varepsilon & 3 & \varepsilon & 1 \\ 2 & \varepsilon & 1 & \varepsilon \\ 1 & 2 & 2 & \varepsilon \\ \varepsilon & \varepsilon & 1 & \varepsilon \end{pmatrix}.$$

Inspecting the communication graph of A in Figure 5.1, it is easily seen that the graph is strongly connected and, consequently, that matrix A is irreducible. We

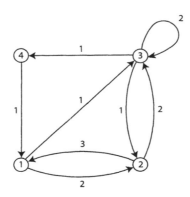

Figure 5.1: Communication graph of Example 5.1.1.

apply Karp's algorithm with $j = 1$, and consider consequently $x(0) = e_1 = (0, \varepsilon, \varepsilon, \varepsilon)^\top$. Note that $n = 4$ and iterating (5.1) four times yields

$$x(1) = \begin{pmatrix} \varepsilon \\ 2 \\ 1 \\ \varepsilon \end{pmatrix}, \quad x(2) = \begin{pmatrix} 5 \\ 2 \\ 4 \\ 2 \end{pmatrix}, \quad x(3) = \begin{pmatrix} 5 \\ 7 \\ 6 \\ 5 \end{pmatrix}, \quad x(4) = \begin{pmatrix} 10 \\ 7 \\ 9 \\ 7 \end{pmatrix}.$$

For $i \in \underline{4}$, the minimization terms $\min_{k=0,\ldots,3} (x_i(4) - x_i(k))/(4 - k)$ read

$$\min\left\{\frac{10}{4}, \infty, \frac{5}{2}, \frac{5}{1}\right\} = 2\frac{1}{2}, \qquad \min\left\{\infty, \frac{5}{3}, \frac{5}{2}, 0\right\} = 0,$$

$$\min\left\{\infty, \frac{8}{3}, \frac{5}{2}, \frac{3}{1}\right\} = 2\frac{1}{2}, \qquad \min\left\{\infty, \infty, \frac{5}{2}, \frac{2}{1}\right\} = 2,$$

respectively, and Karp's algorithm yields

$$\lambda = \max\left\{\frac{5}{2}, 0, 2\right\} = 2\frac{1}{2}$$

for the eigenvalue of A.

In the above example, the matrix A was irreducible. In the next two examples, the applicability of Karp's algorithm will be investigated for matrices that are reducible.

Example 5.1.2 *Let*

$$A = \begin{pmatrix} 1 & 2 & \varepsilon & 7 \\ \varepsilon & 3 & 5 & \varepsilon \\ \varepsilon & 4 & \varepsilon & 3 \\ \varepsilon & 2 & 8 & \varepsilon \end{pmatrix}.$$

The communication graph $\mathcal{G}(A)$ is depicted in Figure 5.2. It is clear that the graph is not strongly connected, as there are no paths from node 1 to the other nodes. Hence, the matrix A is reducible. So Theorem 5.1 does not guarantee that Karp's algorithm yields a correct result. Nevertheless, we are going to apply the algorithm and see where it leads. To apply Karp's algorithm we first take $j = 1$, and consequently, as an initial vector (as in the previous example), $x(0) = e_1$. Repeated application of (5.1) yields that

$$x(k) = \begin{pmatrix} k \\ \varepsilon \\ \varepsilon \\ \varepsilon \end{pmatrix}.$$

For $i = 1$ (and $n = 4$) we have

$$\min_{k=0,\ldots,3} \frac{x_i(4) - x_i(k)}{4 - k} = \min\{1, 1, 1, 1\} = 1,$$

and for $i \in \{2, 3, 4\}$

$$\min_{k=0,\ldots,3} \frac{x_i(4) - x_i(k)}{4 - k} = \min\{0, 0, 0, 0\} = 0,$$

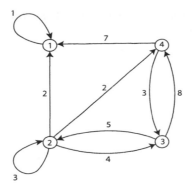

Figure 5.2: Communication graph of Example 5.1.2.

where by convention $\varepsilon - \varepsilon = e$, i.e., $\varepsilon - (-\infty) = 0$; see (2.7). Taking the maximum for $i \in \underline{4}$ results in $\lambda = 1$ as the output of Karp's algorithm. Since $x(1) = A \otimes x(0) = 1 \otimes x(0)$, it follows that $\lambda = 1$ is indeed an eigenvalue of the matrix A with $x(0)$ as an associated eigenvector.

Next, we take $j = 2$, and consequently, as an initial vector $x(0) = e_2 = (\varepsilon, 0, \varepsilon, \varepsilon)^{\top}$. Iterating (5.1) we obtain

$$x(1) = \begin{pmatrix} 2 \\ 3 \\ 4 \\ 2 \end{pmatrix}, \quad x(2) = \begin{pmatrix} 9 \\ 9 \\ 7 \\ 12 \end{pmatrix}, \quad x(3) = \begin{pmatrix} 19 \\ 12 \\ 15 \\ 15 \end{pmatrix}, \quad x(4) = \begin{pmatrix} 22 \\ 20 \\ 18 \\ 23 \end{pmatrix}.$$

For $i \in \underline{4}$, the expression $\min_{k=0,\dots,3} (x_i(4) - x_i(k))/(4 - k)$ gives

$$\min\left\{\infty, \frac{20}{3}, \frac{13}{2}, \frac{3}{1}\right\} = 3, \qquad \min\left\{\frac{20}{4}, \frac{17}{3}, \frac{11}{2}, \frac{8}{1}\right\} = 5,$$

$$\min\left\{\infty, \frac{14}{3}, \frac{11}{2}, \frac{3}{1}\right\} = 3, \qquad \min\left\{\infty, \frac{21}{3}, \frac{11}{2}, \frac{8}{1}\right\} = 5\frac{1}{2},$$

respectively, and Karp's algorithm yields as an output

$$\lambda = 5\frac{1}{2} = \max\left\{3, 5, \frac{11}{2}\right\}.$$

By trial and error it turns out (see also Section 5.2 for an algorithmic approach) that the vector v, given by

$$v = \begin{pmatrix} 24\frac{1}{2} \\ 20 \\ 20\frac{1}{2} \\ 23 \end{pmatrix},$$

satisfies $A \otimes v = \lambda \otimes v$. Hence, λ is indeed an eigenvector of A with associated eigenvector v.

In both cases (i.e., either $j = 1$ or $j = 2, 3, 4$) Karp's algorithm does come up with a correct answer, but the outcome of Karp's algorithm depends on the choice of j.

Example 5.1.3   *Finally, consider a modified version of the previous example in which only the entry a_{11} has been changed:*

$$A = \begin{pmatrix} 6 & 2 & \varepsilon & 7 \\ \varepsilon & 3 & 5 & \varepsilon \\ \varepsilon & 4 & \varepsilon & 3 \\ \varepsilon & 2 & 8 & \varepsilon \end{pmatrix}.$$

As in the previous example, the communication graph of A is not strongly connected. Hence, the matrix A is reducible. Again, in what follows we apply Karp's algorithm even though Theorem 5.1 does not guarantee that the algorithm will yield a correct result.

We first take $j = 1$, and consequently, as an initial vector $x(0) = e_1$. Repeated application of (5.1) yields that

$$x(k) = \begin{pmatrix} 6k \\ \varepsilon \\ \varepsilon \\ \varepsilon \end{pmatrix}.$$

For $i = 1$ we obtain

$$\min_{k=0,\dots,3} \frac{x_i(4) - x_i(k)}{4 - k} = \min\{6, 6, 6, 6\} = 6.$$

As in the previous example, for $i = 2, 3, 4$ the expression is equal to zero. Taking the maximum of the above expressions for $i \in \underline{4}$ yields $\lambda = 6$ as the output of Karp's algorithm. Since it is clear that $x(1) = A \otimes x(0) = 6 \otimes x(0)$, $\lambda = 6$ is indeed an eigenvalue of the matrix A with $x(0)$ as an associated eigenvector.

Now we take $j = 2$, and consequently, as an initial vector $x(0) = e_2$. Iterating, we obtain

$$x(1) = \begin{pmatrix} 2 \\ 3 \\ 4 \\ 2 \end{pmatrix}, \quad x(2) = \begin{pmatrix} 9 \\ 9 \\ 7 \\ 12 \end{pmatrix}, \quad x(3) = \begin{pmatrix} 19 \\ 12 \\ 15 \\ 15 \end{pmatrix}, \quad x(4) = \begin{pmatrix} 25 \\ 20 \\ 18 \\ 23 \end{pmatrix}.$$

For $i \in \underline{4}$, the expression $\min_{k=0,\dots,3} (x_i(4) - x_i(k))/(4 - k)$ is equal to

$$\min\left\{\infty, \frac{23}{3}, \frac{16}{2}, \frac{6}{1}\right\} = 6, \quad \min\left\{\frac{20}{4}, \frac{17}{3}, \frac{11}{2}, \frac{8}{1}\right\} = 5,$$

$$\min\left\{\infty, \frac{14}{3}, \frac{11}{2}, \frac{3}{1}\right\} = 3, \quad \min\left\{\infty, \frac{21}{3}, \frac{11}{2}, \frac{8}{1}\right\} = 5\frac{1}{2},$$

respectively. Taking the maximum of the above expressions for $i \in \underline{4}$ yields again 6 as a candidate for an eigenvalue. It is an eigenvalue since an eigenvector has already been given above.

To conclude this section, we note that Karp's algorithm results in only one value. This value is the eigenvalue of a matrix $A \in \mathbb{R}^{n \times n}_{\max}$ if the matrix is irreducible. If A is reducible, Karp's algorithm also yields an eigenvalue depending on the choice of j. However, the associated eigenvector may have elements equal to ε. Notice that Karp's algorithm provides no means for computing an eigenvector. This drawback of Karp's algorithm is overcome by the power algorithm to be presented in the next section, which simultaneously computes the eigenvalue and a corresponding eigenvector.

5.2 THE POWER ALGORITHM

In this section we assume again that the $n \times n$ matrix A is irreducible. By Theorem 3.9, $x(k)$ following (5.1) will eventually enter a periodic regime. If we evoke Theorem 4.1, then the eigenvalue and a corresponding eigenvector can be computed from such a periodic regime. The resulting algorithm, called the *power algorithm*, is given below.

Algorithm 5.2.1 POWER ALGORITHM

1. *Take an arbitrary initial vector $x(0) = x_0 \neq u[\varepsilon]$; that is, x_0 has at least one finite element.*

2. *Iterate (5.1) until there are integers p, q with $p > q \geq 0$ and a real number c, such that $x(p) = x(q) \otimes c$, i.e., until a periodic regime is reached.*

3. *Compute as the eigenvalue $\lambda = c/(p - q)$ (division in conventional sense).*

4. *Compute as an eigenvector $v = \displaystyle\bigoplus_{j=1}^{p-q} \left(\lambda^{\otimes(p-q-j)} \otimes x(q+j-1) \right).$*

In the following we review the three examples treated in the previous section.

Example 5.2.1   *Reconsider Example 5.1.1. Recall that matrix A is irreducible. Applying the power algorithm, we take as an initial vector $x(0) = e_1$. Iterating (5.1) we obtain (see also Example 5.1.1)*

$$x(1) = \begin{pmatrix} \varepsilon \\ 2 \\ 1 \\ \varepsilon \end{pmatrix}, \quad x(2) = \begin{pmatrix} 5 \\ 2 \\ 4 \\ 2 \end{pmatrix}, \quad x(3) = \begin{pmatrix} 5 \\ 7 \\ 6 \\ 5 \end{pmatrix}, \quad x(4) = \begin{pmatrix} 10 \\ 7 \\ 9 \\ 7 \end{pmatrix}.$$

Note that $x(4) = 5 \otimes x(2)$. In the power algorithm, we therefore have that $p = 4, q = 2$, and $c = 5$, so that consequently $\lambda = 2\frac{1}{2}$. The vector v resulting from the algorithm equals

$$v = (\lambda \otimes x(2)) \oplus x(3) = \begin{pmatrix} 7\frac{1}{2} \\ 4\frac{1}{2} \\ 6\frac{1}{2} \\ 4\frac{1}{2} \end{pmatrix} \oplus \begin{pmatrix} 5 \\ 7 \\ 6 \\ 5 \end{pmatrix} = \begin{pmatrix} 7\frac{1}{2} \\ 7 \\ 6\frac{1}{2} \\ 5 \end{pmatrix}.$$

It is easy to see that indeed

$$A \otimes v = \begin{pmatrix} 10 \\ 9\frac{1}{2} \\ 9 \\ 7\frac{1}{2} \end{pmatrix} = \lambda \otimes v.$$

Hence, v is an eigenvector of matrix A for eigenvalue $\lambda = 2\frac{1}{2}$.

Example 5.2.2   *Reconsider matrix A in Example 5.1.2. Recall that matrix A is reducible, which implies that the power algorithm will thus not necessarily find an eigenvalue and an associated eigenvector.*
    First, we take as an initial vector $x(0) = e_1$. Applying (5.1) we obtain

$$x(1) = \begin{pmatrix} 1 \\ \varepsilon \\ \varepsilon \\ \varepsilon \end{pmatrix} = 1 \otimes \begin{pmatrix} 0 \\ \varepsilon \\ \varepsilon \\ \varepsilon \end{pmatrix} = 1 \otimes x(0),$$

and it immediately follows that $A \otimes v = v \otimes \lambda$ with $\lambda = 1$ and $v = x(0)$.
    Next, we take as an initial vector $x(0) = e_2$. We iterate (5.1), which gives among others

$$x(3) = \begin{pmatrix} 19 \\ 12 \\ 15 \\ 15 \end{pmatrix}, \quad x(4) = \begin{pmatrix} 22 \\ 20 \\ 18 \\ 23 \end{pmatrix}, \quad x(5) = \begin{pmatrix} 30 \\ 23 \\ 26 \\ 26 \end{pmatrix}.$$

It is clear that $x(5) = 11 \otimes x(3)$. Therefore, in the power algorithm we have that $p = 5, q = 3$, and $c = 11$, so that consequently $\lambda = 5\frac{1}{2}$. The vector v resulting from the algorithm equals

$$v = (\lambda \otimes x(3)) \oplus x(4) = \begin{pmatrix} 24\frac{1}{2} \\ 17\frac{1}{2} \\ 20\frac{1}{2} \\ 20\frac{1}{2} \end{pmatrix} \oplus \begin{pmatrix} 22 \\ 20 \\ 18 \\ 23 \end{pmatrix} = \begin{pmatrix} 24\frac{1}{2} \\ 20 \\ 20\frac{1}{2} \\ 23 \end{pmatrix}.$$

It is easy to verify that $A \otimes v = \lambda \otimes v$. Hence, v is an eigenvector of the matrix A for the eigenvalue $\lambda = 5\frac{1}{2}$.
    In the above, the power algorithm does come up with an eigenvalue and an eigenvector, although its outcome depends on the choice of $x(0)$. It is not difficult to see that as long as at least one of the components $x_i(0), i \in \{2, 3, 4\}$, of the initial vector $x(0)$ has a finite value, then the power algorithm will yield the eigenvalue $\lambda = 5\frac{1}{2}$. If $x_i(0) = \varepsilon$ for all $i \in \{2, 3, 4\}$ and $x_1(0)$ is finite, then the algorithm will yield the eigenvalue $\lambda = 1$.
    Note that matrix A has normal form and the eigenvalues of the two diagonal blocks are 1 and $5\frac{1}{2}$, respectively. If we start with a fully finite initial condition, it follows, according to Theorem 3.15, that the cycle-time vector has value $5\frac{1}{2}$ in each of its components. From the above it further follows that $(\mathbf{u}[5\frac{1}{2}], v)$ is a generalized eigenmode of A.

Example 5.2.3 *Reconsider Example 5.1.3. As in the previous example, the matrix A is reducible, and the power algorithm will thus not necessarily yield an eigenvalue and an associated eigenvector of A. If we take as an initial vector $x(0) = e_1$, then we obtain similarly as in the previous example that $x(1) = 6 \otimes x(0)$. Hence, we immediately are in a periodic regime, and it follows that $A \otimes v = \lambda \otimes v$ with $\lambda = 6$ and $v = x(0)$.*

However, if we take as an initial vector $x(0) = e_2$ and iterate, we obtain

$$
\begin{array}{cccc}
x(0) & x(1) & x(2) & x(3) \\[6pt]
\begin{pmatrix} \varepsilon \\ 0 \\ \varepsilon \\ \varepsilon \end{pmatrix} \rightarrow
& \begin{pmatrix} 2 \\ 3 \\ 4 \\ 2 \end{pmatrix} \rightarrow
& \begin{pmatrix} 9 \\ 9 \\ 7 \\ 12 \end{pmatrix} \rightarrow
& \begin{pmatrix} 19 \\ 12 \\ 15 \\ 15 \end{pmatrix} \rightarrow
\end{array}
$$

$$
\begin{array}{cccc}
x(4) & x(5) & x(6) & x(7) \\[6pt]
\begin{pmatrix} 25 \\ 20 \\ 18 \\ 23 \end{pmatrix} \rightarrow
& \begin{pmatrix} 31 \\ 23 \\ 26 \\ 26 \end{pmatrix} \rightarrow
& \begin{pmatrix} 37 \\ 31 \\ 29 \\ 34 \end{pmatrix} \rightarrow
& \begin{pmatrix} 43 \\ 34 \\ 37 \\ 37 \end{pmatrix} \rightarrow
\end{array}
$$

$$
\begin{array}{ccccc}
x(8) & x(9) & x(10) & x(11) & \cdots \\[6pt]
\begin{pmatrix} 49 \\ 42 \\ 40 \\ 45 \end{pmatrix} \rightarrow
& \begin{pmatrix} 55 \\ 45 \\ 48 \\ 48 \end{pmatrix} \rightarrow
& \begin{pmatrix} 61 \\ 53 \\ 51 \\ 56 \end{pmatrix} \rightarrow
& \begin{pmatrix} 67 \\ 56 \\ 59 \\ 59 \end{pmatrix} \rightarrow \cdots .
\end{array}
$$

In the above it is seen that in the long run the first component of $x(k)$ increases each iteration step by 6, while the other components increase on average each iteration step by $5\frac{1}{2}$. Hence, for the chosen initial vector there never can be an overall periodic regime. It turns out that this is always the case for initial vectors $x(0)$ of which at least one of the components $x_i(0), i = 2, 3, 4$, has a finite value. For those initial vectors an overall periodic regime does not exist, and the power algorithm cannot be applied. Again, observe that the above is in correspondence with Theorem 3.15. Indeed, one finds that

$$
\eta = \begin{pmatrix} 6 \\ 5\frac{1}{2} \\ 5\frac{1}{2} \\ 5\frac{1}{2} \end{pmatrix}
$$

is the cycle-time vector when starting with a fully finite initial condition.

To conclude this section we note that if monitoring the iteration process shows that a periodic regime will not be reached, the power algorithm will not terminate. Note that for termination of the algorithm the matrix A does not necessarily have to be irreducible.

5.3 EXERCISES

1. Consider the matrix

$$A = \begin{pmatrix} 5 & 1 \\ 0 & 6 \end{pmatrix}.$$

   Check that matrix A is irreducible, and apply Karp's algorithm to determine the eigenvalue of A.

2. Consider matrix A of exercise 1, and apply the power algorithm to determine the eigenvalue of A and a corresponding eigenvector. Do this starting from

   $$\text{(a)} \quad x(0) = \begin{pmatrix} e \\ \varepsilon \end{pmatrix}, \qquad \text{(b)} \quad x(0) = \begin{pmatrix} \varepsilon \\ e \end{pmatrix}.$$

3. Combine Karp's algorithm and the power algorithm to obtain an algorithm that in as few as possible iterations of (5.1) results in the eigenvalue of an irreducible matrix A.

4. Explain why in Example 5.2.2 the power algorithm comes up with the eigenvalue $5\frac{1}{2}$ if at least one of the components $x_i(0), i = 2, 3, 4$, has a finite value.

5. Eplain why in Example 5.2.3 the power algorithm does not work if at least one of the components $x_i(0), i = 2, 3, 4$, has a finite value.

5.4 NOTES

Example 5.1.1 is due to [17], and Examples 5.1.2 and 5.1.3 are based upon [23]. Section 5.2 is based on [83]. Exercise 2 is inspired by [81].

The complexity of Karp's algorithm is of order n^3, while the complexity of the power algorithm is less clear. In particular, the length of the transient behavior (i.e., the number of steps to reach the periodic regime) can be large depending on the value of the entries of the matrix. Only very conservative upper bounds exist for the length of the transition behavior. See, for instance, [50] or [81].

Numerical methods, other than presented in this chapter and the next, have been proposed in the literature, based, for example, on linear programming techniques or on finding the root of the characteristic polynomial in max-plus algebra; see [5] and [75]. The ones presented in the current book turned out to be the most powerful.

Chapter Six

A Numerical Procedure for Eigenvalues of Reducible Matrices

The generalized eigenmode of a square matrix has been introduced and studied in Chapter 3. More specifically, in Sections 3.2 and 3.3 the existence of a generalized eigenmode of a square regular matrix has been proved by making use of its normal form. As the proofs in Sections 3.2 and 3.3 are constructive, a conceptual algorithm has been obtained by which a generalized eigenmode in principle can be computed. See in particular the proof of Corollary 3.16. However, the obtained algorithm heavily relies on a normal form of the matrix involved.

In this chapter an alternative algorithm is presented. *Howard's algorithm*, as it is called, is used to compute the generalized eigenmode of a square regular matrix in a direct way, avoiding its normal form. Despite the differences, many similarities exist between the approaches and algorithms in Chapter 3 and this chapter. The notion of a bouquet matrix, studied in Section 3.4, will play a prominent role in the current chapter.

As noted, the existence of a generalized eigenmode of a square regular matrix has been treated in Sections 3.2 and 3.3, where it has been shown that for any regular matrix $A \in \mathbb{R}_{\max}^{n \times n}$, finite vectors $\eta, v \in \mathbb{R}^n$ exist such that for all $k \geq 0$

$$A \otimes (v + k \times \eta) = v + (k + 1) \times \eta. \tag{6.1}$$

Note that if η and v satisfy (6.1) for all $k \geq 0$, then for each k, the maximum of $a_{ij} + v_j + k\eta_j, j \in \underline{n}$, in the matrix product on the left-hand side of (6.1), needs to be attained for just one suitable pair (j, i) per equation. Therefore, a first (brute force) attempt to determine a solution of (6.1) is to assign, independently of k, to each $i \in \underline{n}$ precisely one j from the set $\underline{n}$ for which the maximum in the ith equation of (6.1) might be attained. Then, with each $i \in \underline{n}$, precisely one j from the set $\underline{n}$ is associated. This type of association is called *policy* in the literature.

A policy Π can be seen as a mapping $\Pi : \mathcal{N}(A) \to \mathcal{D}(A)$, assigning to each node $i \in \mathcal{N}(A)$ an arc $\Pi_i \in \mathcal{D}(A)$ such that node i is the end node of arc Π_i. Let A^Π be the matrix obtained from A by keeping all the entries of A corresponding to the arcs $\Pi_1, \Pi_2, \ldots, \Pi_n$ and by replacing all the other entries of A by ε. The matrix A^Π is often referred to as a *policy matrix*. Note that a policy matrix has precisely one finite element in each row. Examples of policy matrices have therefore already been encountered in Section 3.4 in the form of sunflower and bouquet matrices. In fact, policy matrices and bouquet matrices are different names for the same type of matrices.

In Section 3.4, we gave a graph-theoretical method to compute a generalized eigenmode of the policy/bouquet matrix A^Π. The method can be implemented efficiently. The problem of finding a generalized eigenmode of A can therefore be reduced to that of finding a policy Π such that the generalized eigenmode of A^Π is also a generalized eigenmode of the overall matrix A. The solution of (6.1) is attained by at least one policy, and since there are only finitely many policies, it is clear that searching the space of policies will eventually lead to the solution of (6.1). Unfortunately, for a fully finite matrix A the number of policies is equal to n^n, indicating that in general the number of policies grows exponentially with n. Hence, a clever method has to be derived for improving a policy when it is not yet the right one.

Howard's algorithm, also known as the *policy iteration algorithm*, is an iterative algorithm for computing a generalized eigenmode. First, a policy is chosen and the eigenmode of the corresponding policy matrix is computed. This part of the overall algorithm is called *value determination* and is presented in algorithmic form in Section 6.1.1. Next, it is tested whether the generalized eigenmode of the policy matrix is already a generalized eigenmode of the original matrix. If so, a solution of (6.1) for all $k \geq 0$ has been found. If not, the policy has to be adapted. To obtain a clever adaptation scheme, first the equations for η and v in (6.1), which must hold for all $k \geq 0$, will be replaced by an equivalent set of equations for η and v that are independent of k. Based on the equivalent set of equations, a scheme will be derived on how to improve the chosen but incorrect policy. This scheme, called *policy improvement*, will be presented in algorithmic form in Section 6.1.2.

Howard's algorithm will be presented in Section 6.1.3. It is based on the algorithms developed and presented in Sections 6.1.1 and 6.1.2. Numerical examples are provided in Section 6.2.

6.1 HOWARD'S ALGORITHM

6.1.1 Value determination

Consider the matrix A, and let Π be a given policy. Our aim is to compute a generalized eigenmode of the policy matrix A^Π; that is, we wish to compute vectors $\eta, v \in \mathbb{R}^n$ such that for all $k \geq 0$

$$A^\Pi \otimes (v + k \times \eta) = v + (k + 1) \times \eta.$$

The matrix A^Π is a bouquet matrix, and the communication graph $\mathcal{G}(A^\Pi)$ is made up of one or more subgraphs, being sunflower graphs, with the associated matrices, being sunflower matrices. In Section 3.4 it has been shown that the eigenvalue of a sunflower matrix exists and equals the mean of the (only) circuit in the associated sunflower graph. A corresponding eigenvector follows by going along the circuit and paths, starting from an arbitrary chosen node in the circuit. Combining the obtained eigenvalues and eigenvectors, a generalized eigenmode of the matrix A^Π is obtained; see (3.15) and (3.16) in Section 3.4 for details.

The algorithm in Section 3.4 is going to be part of Howard's algorithm and as such is applied to compute a generalized eigenmode of a policy (matrix). Note that for each obtained eigenmode, the cycle-time vector η will be unique, whereas the vector v will not be unique. Indeed, v may be changed by increasing each of its components (belonging to the same sunflower) with the same number.

In order to ensure that Howard's algorithm terminates, the new value of the vector v is partly based on its old value, thereby introducing a kind of ordering between the eigenmodes corresponding to subsequent policies. More specifically, in each sunflower graph a node is chosen in the associated (unique) circuit, and the corresponding component of v is kept at the value it had obtained previously. The components corresponding to the other nodes of the sunflower are given newly computed values. When starting Howard's algorithm, the vector v is set to the unit vector, i.e., $v := \mathbf{u}$. In the algorithm stated below, the direct predecessor of node j in the communication graph of A^Π is denoted by $\pi_\Pi(j)$.

Algorithm 6.1.1 VALUE DETERMINATION

1. *Find a circuit ζ in the graph $\mathcal{G}(A^\Pi)$.*

2. *Compute the average weight of ζ and denote it by $\bar{\eta}_\zeta$.*

3. *Select a node in the circuit ζ, say, node j, and set $\eta_j := \bar{\eta}_\zeta$. Further, set $v_j := v_j$, i.e., let v_j keep the value it had obtained previously.*

4. *Visit all nodes in the graph $\mathcal{G}(A^\Pi)$ that are reachable from node j. If node i is visited in this process, then set $\eta_i := \bar{\eta}_\zeta$ and calculate v_i from the iteration $v_i = a_{i,\pi_\Pi(i)} - \bar{\eta}_\zeta + v_{\pi_\Pi(i)}$.*

5. *If there remain nodes that are not reachable from node j in $\mathcal{G}(A^\Pi)$, restart the algorithm with the steps 1 to 4 for the graph made up of these nodes and the associated arcs from $\mathcal{G}(A^\Pi)$.*

The reason for dealing with the value of v_j as in step 3 is that monotonicity is then achieved with respect to the following lexicographical ordering. Given two candidate generalized eigenmodes (η, v) and (η', v'), with $\eta, v, \eta', v' \in \mathbb{R}^n$, it is said that $(\eta, v) \leq (\eta', v')$ if either $\eta \leq \eta'$ or $\eta = \eta'$ and $v \leq v'$. Here $\eta \leq \eta'$ means that η componentwise is less than or equal to η', with at least one strict inequality, and similarly for $v \leq v'$. This monotonicity can be used to prove (see Section 6.5) that the overall algorithm converges.

Given a policy Π and a generalized eigenmode (η, v) of the associated policy matrix A^Π, it can be checked whether (η, v) is also a generalized eigenmode of the matrix A itself. If so, we have found two finite vectors η and v such that (6.1) is satisfied for all $k \geq 0$ and an eigenmode has been found. If not, then the policy Π is not yet a correct policy for matrix A and should be replaced by finding a new (improved) policy Π'. How this is done is explained in the policy improvement algorithm in Section 6.1.2.

6.1.2 Policy improvement

As announced, in this section first the equations in (6.1), which must hold for all $k \geq 0$, will be shown to be equivalent to two sets of alternative equations that are independent of k. For that purpose, consider a regular matrix $A \in \mathbb{R}_{max}^{n \times n}$. If $\eta, v \in \mathbb{R}^n$ satisfy (6.1) for all $k \geq 0$, this means in conventional notation that η and v are such that for all $k \geq 0$ and all $i \in \underline{n}$

$$\max_{j \in \underline{n}} \{a_{ij} + v_j + k\eta_j\} = v_i + (k+1)\eta_i. \tag{6.2}$$

Note that for notational convenience the multiplication sign $\times$ is omitted in (6.2). From the definition of the arc set, it follows that $a_{ij} \neq \varepsilon$ if and only if $(j, i) \in \mathcal{D}(A)$. Therefore, the previous implies that for all $k \geq 0$ and all $i \in \underline{n}$

$$\max_{j \in \mathcal{D}(A)_i} \{a_{ij} + v_j + k\eta_j\} = v_i + (k+1)\eta_i,$$

where $\mathcal{D}(A)_i = \{j : (j, i) \in \mathcal{D}(A)\}$. Note that, for a fixed node i, $j \in \mathcal{D}(A)_i$ if and only if $(j, i) \in \mathcal{D}(A)$. Hence, the above maximization over nodes in $\mathcal{D}(A)_i$ can also be written as a maximization over arcs in $\mathcal{D}(A)$ that go to the specific node i. Therefore, it follows from (6.2) that for all $k \geq 0$ and all $i \in \underline{n}$

$$\max_{(j,i) \in \mathcal{D}(A)} \{a_{ij} + v_j + k\eta_j\} = v_i + (k+1)\eta_i. \tag{6.3}$$

In the following the above notation with maximization over an arc set will used extensively.

Dividing both sides in (6.3) by $k > 0$ yields for all $i \in \underline{n}$

$$\max_{(j,i) \in \mathcal{D}(A)} \left\{ \frac{a_{ij} + v_j + k\eta_j}{k} \right\} = \frac{v_i + (k+1)\eta_i}{k}.$$

Note that the components of both η and v are finite. Also all the entries a_{ij} with $(j, i) \in \mathcal{D}(A)$ are finite. Hence, in the last equality the limit exists for k to ∞ and satisfies for all $i \in \underline{n}$

$$\max_{(j,i) \in \mathcal{D}(A)} \eta_j = \eta_i. \tag{6.4}$$

Given a finite vector η satisfying the above equations, define the matrix $\bar{A}$ as

$$[\bar{A}]_{ij} \overset{\text{def}}{=} \begin{cases} [A]_{ij} & \text{if } (j, i) \in \mathcal{D}(A) \text{ and } \eta_i = \eta_j, \\ \varepsilon & \text{otherwise.} \end{cases} \tag{6.5}$$

Note that due to the regularity of A the arc set corresponding to matrix $\bar{A}$ is not empty; see exercise 1. Clearly, this arc set is given by

$$\mathcal{D}(\bar{A}) = \{(j, i) \in \mathcal{D}(A) | \eta_i = \eta_j\}. \tag{6.6}$$

With (6.4) and (6.6) it is clear that for the above η

$$\begin{cases} \eta_i = \eta_j & \text{for all } (j, i) \in \mathcal{D}(\bar{A}), \\ \eta_i > \eta_j & \text{for all } (j, i) \in \mathcal{D}(A) \backslash \mathcal{D}(\bar{A}). \end{cases}$$

Now it follows that for k large enough and for any $i \in \underline{n}$ the expression

$$\max_{(j,i) \in \mathcal{D}(A)} \{a_{ij} + v_j + k\eta_j\} = v_i + (k+1)\eta_i$$

can be replaced by

$$\max_{(j,i) \in \mathcal{D}(\bar{A})} \{a_{ij} + v_j + k\eta_j\} = v_i + (k+1)\eta_i. \tag{6.7}$$

Indeed, it is easy to see that since $\eta_i > \eta_j$ for all $(j,i) \in \mathcal{D}(A) \backslash \mathcal{D}(\bar{A})$, there exists an integer $K \geq 0$ such that for all $k \geq K$ and any $i \in \underline{n}$

$$\max_{(j,i)\in\mathcal{D}(A)\backslash\mathcal{D}(\bar{A})} \{a_{ij} + v_j + k\eta_j\} \leq v_i + (k+1)\eta_i. \tag{6.8}$$

The reason for this is that the left-hand side in (6.8) grows at a rate of at most $\max\{\eta_j | (j,i) \in \mathcal{D}(A)\backslash\mathcal{D}(\bar{A})\}$, which is less than η_i, the rate at which the right-hand side grows.

Further, because $\eta_i = \eta_j$ for all $(j,i) \in \mathcal{D}(\bar{A})$, it follows now from (6.7) that for all $i \in \underline{n}$

$$\max_{(j,i)\in\mathcal{D}(\bar{A})} \{a_{ij} + v_j - \eta_j\} = v_i.$$

Hence, it follows from (6.2), with $\mathcal{D}(\bar{A})$ defined in (6.6), that for all $i \in \underline{n}$

$$\max_{(j,i)\in\mathcal{D}(A)} \eta_j = \eta_i, \tag{6.9}$$

$$\max_{(j,i)\in\mathcal{D}(\bar{A})} \{a_{ij} + v_j - \eta_j\} = v_i. \tag{6.10}$$

Conversely, given finite vectors η and v that satisfy (6.9) and (6.10) for all $i \in \underline{n}$, it follows directly for all $k \geq 0$ and all $i \in \underline{n}$ that

$$v_i + (k+1)\eta_i \stackrel{(6.10)}{=} \max_{(j,i)\in\mathcal{D}(\bar{A})} \{a_{ij} + v_j - \eta_j\} + (k+1)\eta_i$$

$$= \max_{(j,i)\in\mathcal{D}(\bar{A})} \{a_{ij} + v_j - \eta_j + (k+1)\eta_i\}$$

$$\stackrel{(6.6)}{=} \max_{(j,i)\in\mathcal{D}(\bar{A})} \{a_{ij} + v_j + k\eta_j\}$$

$$\leq \max_{(j,i)\in\mathcal{D}(A)} \{a_{ij} + v_j + k\eta_j\}. \tag{6.11}$$

Again by (6.8) it follows easily that for k large enough the inequality in (6.11) has to be an equality.

Hence, finding finite vectors η and v such that (6.9) and (6.10) are satisfied for all $i \in \underline{n}$, with $\mathcal{D}(\bar{A})$ defined in (6.6), is equivalent to finding finite vectors η and v such that (6.1) is satisfied for k large enough, say, for $k \geq K$, with K large enough. Then, by redefining $v := v + K \times \eta$, it is easy to see that the latter is also equivalent to finding finite vectors η and v such that (6.1) is satisfied for all $k \geq 0$.

Now using equations (6.9) and (6.10) it can be checked if a solution of the restricted problem, corresponding to a chosen policy, is a solution of the original problem. If not, the equations also can be used to obtain an improved policy. All this is explained below, where the starting point is a policy Π and two associated vectors η and v such that $A^{\Pi} \otimes (v + k \times \eta) = v + (k+1) \times \eta$, for all $k \geq 0$, or, equivalently, such that for all $i \in \underline{n}$

$$\max_{(j,i)\in\mathcal{D}(A^{\Pi})} \eta_j = \eta_i, \qquad \max_{(j,i)\in\mathcal{D}(A^{\Pi})} \{a_{ij} + v_j - \eta_j\} = v_i.$$

The idea is now to check whether the vectors η and v are also such that for all $i \in \underline{n}$

$$\max_{(j,i)\in\mathcal{D}(\mathcal{A})} \eta_j = \eta_i, \qquad \max_{(j,i)\in\mathcal{D}(\bar{A})} \{a_{ij} + v_j - \eta_j\} = v_i$$

and, if not, to give a better policy. This all can be done by the two steps described next.

The first step is to check whether η satisfies (6.9) (i.e., to check whether it holds that $\max\{\eta_j | (j, i) \in \mathcal{D}(A)\} = \eta_i$ for all $i \in \underline{n}$).

1. Determine the set

$$I_1 = \left\{ i \in \mathcal{N}(A) \,\middle|\, \eta_i < \max_{(j,i) \in \mathcal{D}(A)} \eta_j \right\}.$$

If $I_1 \neq \emptyset$, then determine for all $i \in I_1$ the sets

$$\mathcal{D}(A)_i^1 = \left\{ (k, i) \in \mathcal{D}(A) \,\middle|\, \eta_k = \max_{(j,i) \in \mathcal{D}(A)} \eta_j \right\}.$$

If $I_1 \neq \emptyset$, then the present η and the corresponding policy Π are not yet correct. A better policy Π' has to be found. This is done by adjusting Π for those nodes i that do not satisfy (6.9), i.e., for nodes in the set I_1. This can be done as follows. Define

$$\Pi_i' := \begin{cases} (k, i) & \text{for some } (k, i) \in \mathcal{D}(A)_i^1 \text{ if } i \in I_1, \\ \Pi_i & \text{if } i \notin I_1. \end{cases}$$

Next, return to the value determination in Section 6.1.1 to compute a new generalized eigenmode for the policy matrix $A^{\Pi'}$.

If $I_1 = \emptyset$, then the present η is such that (6.9) is satisfied for all $i \in \underline{n}$. Next, (6.10) has to be considered. Therefore, consider $\mathcal{D}(\bar{A})$ as defined in (6.6) and continue as follows.

2. Determine the set

$$I_2 = \left\{ i \in \mathcal{N}(A) \,\middle|\, v_i < \max_{(j,i) \in \mathcal{D}(\bar{A})} (a_{ij} + v_j - \eta_j) \right\}.$$

If $I_2 \neq \emptyset$, then determine for all $i \in I_2$ the sets

$$\mathcal{D}(A)_i^2 = \left\{ (k, i) \in D(A) \,\middle|\, a_{ik} + v_k - \eta_k = \max_{(j,i) \in \mathcal{D}(\bar{A})} (a_{ij} + v_j - \eta_j) \right\}.$$

With this second step it is checked whether the vectors η and v satisfy (6.10); that is, it is checked whether $\max\{a_{ij} + v_j - \eta_j | (j, i) \in \mathcal{D}(\bar{A})\} = v_i$ for all $i \in \underline{n}$.

If $I_2 \neq \emptyset$, then the current η and v, and consequently the current policy, are not yet correct. A new policy Π' is found by adjusting Π for those nodes that do not satisfy (6.10), i.e., for nodes in the set I_2. This can be done as follows. Define

$$\Pi_i' := \begin{cases} (k, i) & \text{for some } (k, i) \in \mathcal{D}(A)_i^2 \text{ if } i \in I_2, \\ \Pi_i & \text{if } i \notin I_2. \end{cases}$$

Next, return to the value determination in Section 6.1.1 to compute a new generalized eigenmode for the policy matrix $A^{\Pi'}$.

If $I_2 = \emptyset$, then the current η and v are such that (6.10) is satisfied for all $i \in \underline{n}$ and the algorithm terminates.

The above steps are summarized in the following algorithm.

Algorithm 6.1.2 POLICY IMPROVEMENT

1. *Determine the set* $I_1 = \left\{ i \in \mathcal{N}(A) \,\middle|\, \eta_i < \max_{(j,i)\in\mathcal{D}(A)} \eta_j \right\}.$

   - *If* $I_1 = \emptyset$, *the current* η *is such that (6.9) is satisfied for all* $i \in \underline{n}$. *Then compute* $\mathcal{D}(\bar{A}) = \{(j,i) \in \mathcal{D}(A) | \eta_i = \eta_j\}$ *and continue with step 2.*

   - *If* $I_1 \neq \emptyset$, *then determine for all* $i \in I_1$ *the sets*

   $$\mathcal{D}(A)_i^1 = \left\{ (k,i) \in \mathcal{D}(A) \,\middle|\, \eta_k = \max_{(j,i)\in\mathcal{D}(A)} \eta_j \right\}.$$

   Define a new policy Π' *as*

   $$\Pi_i' := \begin{cases} (k,i) & \text{for some } (k,i) \in \mathcal{D}(A)_i^1 \text{ if } i \in I_1, \\ \Pi_i & \text{if } i \notin I_1. \end{cases}$$

2. *Determine the set* $I_2 = \left\{ i \in \mathcal{N}(A) \,\middle|\, v_i < \max_{(j,i)\in\mathcal{D}(\bar{A})} (a_{ij} + v_j - \eta_j) \right\}.$

   - *If* $I_2 = \emptyset$, *then the current* η *and* v *are such that both (6.9) and (6.10) are satisfied for all* $i \in \underline{n}$. *Hence, the algorithm can stop.*

   - *If* $I_2 \neq \emptyset$, *then determine for all* $i \in I_2$ *the sets*

   $$\mathcal{D}(A)_i^2 = \left\{ (k,i) \in D(A) \,\middle|\, a_{ik} + v_k - \eta_k = \max_{(j,i)\in\mathcal{D}(\bar{A})} (a_{ij} + v_j - \eta_j) \right\}.$$

   Define a new policy Π' *as*

   $$\Pi_i' := \begin{cases} (k,i) & \text{for some } (k,i) \in \mathcal{D}(A)_i^2 \text{ if } i \in I_2, \\ \Pi_i & \text{if } i \notin I_2. \end{cases}$$

6.1.3 The overall algorithm

In this section we put together the ideas presented in the previous sections. This results in an algorithm that yields a generalized eigenmode (η, v) for square regular matrices.

Algorithm 6.1.3 HOWARD'S ALGORITHM

Choose an arbitrary policy $\Pi(1)$ *and set* $v^{\Pi(0)} := \mathbf{u}$. *Apply the following steps for* $k \geq 1$.

1. *Compute a generalized eigenmode* $(\eta^{\Pi(k)}, v^{\Pi(k)})$ *for* $A^{\Pi(k)}$ *via the value determination algorithm, i.e., Algorithm 6.1.1, where* $v_j^{\Pi(k)} := v_j^{\Pi(k-1)}$ *for selected nodes* j, *one in each of the circuits of* $\mathcal{G}(A^{\Pi(k)})$.

2. *Check whether* $(\eta^{\Pi(k)}, v^{\Pi(k)})$ *is also a generalized eigenmode of* A. *If so, the algorithm terminates. Otherwise, construct a new policy* $\Pi(k + 1)$ *via the policy improvement algorithm, i.e., Algorithm 6.1.2. Set* $k := k + 1$, *and return to step 1.*

Howard's algorithm is a very efficient algorithm for computing the generalized eigenmode of a regular matrix A. It has been reported to be on average almost linear in time; that is, in processing time, the algorithm is on average linearly proportional to the number of finite entries in the matrix A, though a tight upper bound has not yet been found. It is also conjectured that in the worst case the run time is polynomial in the number of finite entries of A. For more details on this, see [23].

6.2 EXAMPLES

In this section Howard's algorithm will be illustrated by means of the matrices in Examples 5.1.2 and 5.1.3.

Example 6.2.1   *Consider the matrix A studied in Example 5.1.2, given by*

$$A = \begin{pmatrix} 1 & 2 & \varepsilon & 7 \\ \varepsilon & 3 & 5 & \varepsilon \\ \varepsilon & 4 & \varepsilon & 3 \\ \varepsilon & 2 & 8 & \varepsilon \end{pmatrix}.$$

The communication graph of A is depicted in Figure 5.2. Recall that the graph is not strongly connected and that A is reducible.

Take as a starting point the policy $\Pi(1) = \{(1,1),(2,2),(2,3),(2,4)\}$ and $v^{\Pi(0)} := \mathbf{u}$. The graph $\mathcal{G}(A^{\Pi(1)})$ associated to $A^{\Pi(1)}$ is depicted in Figure 6.1. To obtain a generalized eigenmode of $A^{\Pi(1)}$, step 1 of Howard's algorithm will

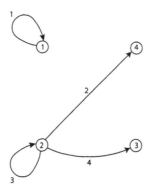

Figure 6.1: Communication graph of $A^{\Pi(1)}$.

be applied (i.e., the value determination according to Algorithm 6.1.1). Note that $\mathcal{G}(A^{\Pi(1)})$ contains two circuits, corresponding to two sunflowers. One circuit, denoted $\zeta_1^{\Pi(1)}$, is given by the arc $(1,1)$, with average weight 1. Clearly, the circuit consists of just one node, namely, node 1. Choose this node, and set $\eta_1^{\Pi(1)} := 1$ and $v_1^{\Pi(1)} := v_1^{\Pi(0)}(= 0)$. As there are no other vertices that can be reached from node 1 in $\mathcal{G}(A^{\Pi(1)})$, step 1 through 4 of Algorithm 6.1.1 are now completed for the circuit $\zeta_1^{\Pi(1)}$.

Next, steps 1 through 5 of Algorithm 6.1.1 have to be repeated for the remaining nodes and arcs of $\mathcal{G}(A^{\Pi(1)})$. The second circuit, denoted $\zeta_2^{\Pi(1)}$, is given by the arc $(2,2)$ and has average weight 3. This circuit consists of just one node, namely, node 2. Choose this node, and set $\eta_2^{\Pi(1)} := 3$ and $v_2^{\Pi(1)} := v_2^{\Pi(0)}(= 0)$. Now note that there are vertices that can be reached from node 2 in $\mathcal{G}(A^{\Pi(1)})$, namely, nodes 3 and 4. Therefore, set $\eta_3^{\Pi(1)} := 3$ and $\eta_4^{\Pi(1)} := 3$. Further, compute $v_3^{\Pi(1)}$ and $v_4^{\Pi(1)}$ according to step 4 of Algorithm 6.1.1. Specifically, $\pi_{\Pi(1)}(3) = 2$ and $\pi_{\Pi(1)}(4) = 2$, which gives

$$v_3^{\Pi(1)} = a_{32} + v_2^{\Pi(1)} - \eta_2^{\Pi(1)} = 1$$

and

$$v_4^{\Pi(1)} = a_{42} + v_2^{\Pi(1)} - \eta_2^{\Pi(1)} = -1.$$

Since all nodes have been treated, a generalized eigenmode $(\eta^{\Pi(1)}, v^{\Pi(1)})$ of $A^{\Pi(1)}$ is obtained with

$$\eta^{\Pi(1)} = \begin{pmatrix} 1 \\ 3 \\ 3 \\ 3 \end{pmatrix} \quad \text{and} \quad v^{\Pi(1)} = \begin{pmatrix} 0 \\ 0 \\ 1 \\ -1 \end{pmatrix}.$$

Next it has to be checked whether $(\eta^{\Pi(1)}, v^{\Pi(1)})$ is also an eigenmode of A itself, i.e., whether equations (6.9) and (6.10) are satisfied. This will be done by applying step 2 of Howard's algorithm (i.e., the policy improvement according to Algorithm 6.1.2). First, the set I_1 has to be determined for $(\eta^{\Pi(1)}, v^{\Pi(1)})$ as given above. It follows here easily that $I_1 = \{1\}$. The corresponding arc set $\mathcal{D}(A)_1^1$ is equal to $\{(2,1),(4,1)\}$. Using the improvement as in step 1 of Algorithm 6.1.2, it follows that, for instance, $\Pi(2) = \{(2,1),(2,2),(2,3),(2,4)\}$ can be taken; that is, to obtain $\Pi(2)$ from $\Pi(1)$, just replace $(1,1)$ by $(2,1)$ and keep the other arcs. The graph $\mathcal{G}(A^{\Pi(2)})$ associated to $A^{\Pi(2)}$ is depicted in Figure 6.2. To obtain a gen-

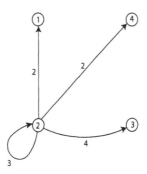

Figure 6.2: Communication graph of $A^{\Pi(2)}$.

eralized eigenmode of $A^{\Pi(2)}$, steps 1 through 5 of Algorithm 6.1.1 will be applied. Note that the graph $\mathcal{G}(A^{\Pi(2)})$ contains just one circuit given by the arc $(2,2)$. Denote this circuit by $\zeta^{\Pi(2)}$. Its average weight is 3, and it consists of just one node, *namely, node 2. Choose this node, and set $\eta_2^{\Pi(2)} := 3$ and $v_2^{\Pi(2)} := v_2^{\Pi(1)}(= 0)$. Now note that all the vertices can be reached from node 2 in $\mathcal{G}(A^{\Pi(2)})$. Therefore, set $\eta_1^{\Pi(2)} := 3$, $\eta_3^{\Pi(2)} := 3$, and $\eta_4^{\Pi(2)} := 3$. Note that $\pi_{\Pi(2)}(1) = 2$, $\pi_{\Pi(2)}(3) = 2$, and $\pi_{\Pi(2)}(4) = 2$. Hence, it follows that*

$$v_1^{\Pi(2)} = a_{12} + v_2^{\Pi(2)} - \eta_2^{\Pi(2)} = -1,$$

$$v_3^{\Pi(2)} = a_{32} + v_2^{\Pi(2)} - \eta_2^{\Pi(2)} = 1,$$

and

$$v_4^{\Pi(2)} = a_{42} + v_2^{\Pi(2)} - \eta_2^{\Pi(2)} = -1,$$

which gives a generalized eigenmode $(\eta^{\Pi(2)}, v^{\Pi(2)})$ of $A^{\Pi(2)}$ with

$$\eta^{\Pi(2)} = \begin{pmatrix} 3 \\ 3 \\ 3 \\ 3 \end{pmatrix} \quad \text{and} \quad v^{\Pi(2)} = \begin{pmatrix} -1 \\ 0 \\ 1 \\ -1 \end{pmatrix}.$$

It can be checked whether $(\eta^{\Pi(2)}, v^{\Pi(2)})$ is also an eigenmode of A itself by applying steps 1 and 2 of Algorithm 6.1.2 again. In order to do so, first, set I_1 has to be determined for $(\eta^{\Pi(2)}, v^{\Pi(2)})$ as given above. Since all components of $\eta^{\Pi(2)}$ have the same value, it follows that $I_1 = \emptyset$. Also it follows that $\mathcal{D}(\bar{A})$, defined in (6.6), coincides with $\mathcal{D}(A)$. Hence, step 1 can be skipped and step 2 has to be done. So, set I_2 has to be determined. Therefore, determine the vector $w^{\Pi(2)}$, with components $w_i^{\Pi(2)} = \max\limits_{(j,i)\in\mathcal{D}(\bar{A})} \{a_{ij} + v_j^{\Pi(2)} - \eta_j^{\Pi(2)}\}$ for all $i \in \underline{4}$, resulting in

$$w^{\Pi(2)} = \begin{pmatrix} 3 \\ 3 \\ 1 \\ 6 \end{pmatrix}.$$

Comparing $w^{\Pi(2)}$ with $v^{\Pi(2)}$, as done in step 2, it is clear that $I_2 = \{1, 2, 4\}$. The corresponding arc sets $\mathcal{D}(A)_i^2$ are given by $\mathcal{D}(A)_1^2 = \{(4, 1)\}$, $\mathcal{D}(A)_2^2 = \{(3, 2)\}$, and $\mathcal{D}(A)_4^2 = \{(3, 4)\}$. Using the improvement as in step 2, it follows that $\Pi(3) = \{(4, 1), (3, 2), (2, 3), (3, 4)\}$ has to be taken. The graph $\mathcal{G}(A^{\Pi(3)})$ associated to $A^{\Pi(3)}$ is depicted in Figure 6.3. Again steps 1 through 5 of Algorithm 6.1.1 will be applied to obtain a generalized eigenmode of $A^{\Pi(3)}$. Note that the graph $\mathcal{G}(A^{\Pi(3)})$ contains just one circuit given by the arcs $\{(2, 3), (3, 2)\}$, with average weight equal to $4\frac{1}{2}$. Choose node 2 on this circuit. Performing steps 3 and 4 of Algorithm 6.1.1 yields a generalized eigenmode $(\eta^{\Pi(3)}, v^{\Pi(3)})$ of $A^{\Pi(3)}$, with

$$\eta^{\Pi(3)} = \begin{pmatrix} 4\frac{1}{2} \\ 4\frac{1}{2} \\ 4\frac{1}{2} \\ 4\frac{1}{2} \end{pmatrix} \quad \text{and} \quad v^{\Pi(3)} = \begin{pmatrix} 5\frac{1}{2} \\ 0 \\ -\frac{1}{2} \\ 3 \end{pmatrix}.$$

It can be checked whether $(\eta^{\Pi(3)}, v^{\Pi(3)})$ is also an eigenmode of A itself by applying steps 1 and 2 of Algorithm 6.1.2. Therefore, the set I_1 and possibly set I_2

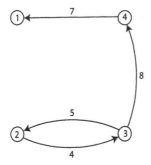

Figure 6.3: Communication graph of $A^{\Pi(3)}$.

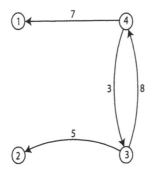

Figure 6.4: Communication graph of $A^{\Pi(4)}$.

have to be determined. It turns out that $I_1 = \emptyset$ and that $I_2 = \{3\}$. The corresponding arc set $\mathcal{D}(A)_3^2$ is given by $\mathcal{D}(A)_3^2 = \{(4,3)\}$. Using the improvement as in step 2, it follows that $\Pi(4) = \{(4,1),(3,2),(4,3),(3,4)\}$ has to be taken. The graph $\mathcal{G}(A^{\Pi(4)})$ associated to $A^{\Pi(4)}$ is depicted in Figure 6.4. Again steps 1 through 5 of Algorithm 6.1.1 will be applied. Note that the graph $\mathcal{G}(A^{\Pi(4)})$ contains just one circuit given by the arcs $\{(3,4),(4,3)\}$, with average weight being $5\frac{1}{2}$. Pick node 3 on this circuit. Performing steps 3 and 4 of Algorithm 6.1.1 results in a generalized eigenmode $(\eta^{\Pi(4)}, v^{\Pi(4)})$ of $A^{\Pi(4)}$, with

$$\eta^{\Pi(4)} = \begin{pmatrix} 5\frac{1}{2} \\ 5\frac{1}{2} \\ 5\frac{1}{2} \\ 5\frac{1}{2} \end{pmatrix} \qquad \text{and} \qquad v^{\Pi(4)} = \begin{pmatrix} 3\frac{1}{2} \\ -1 \\ -\frac{1}{2} \\ 2 \end{pmatrix}.$$

It can be checked whether $(\eta^{\Pi(4)}, v^{\Pi(4)})$ is also an eigenmode of A itself by applying steps 1 and 2 of Algorithm 6.1.2. It turns out that both $I_1 = \emptyset$ and $I_2 = \emptyset$. Hence, the algorithm can stop and the pair $(\eta^{\Pi(4)}, v^{\Pi(4)})$ satisfies (6.9) and (6.10), for all $i \in \underline{4}$, so that

$$A \otimes (v^{\Pi(4)} + k \times \eta^{\Pi(4)}) = v^{\Pi(4)} + (k+1) \times \eta^{\Pi(4)},$$

for all $k \geq K$, with $K \geq 0$ large enough. By simple verification it follows that here $K = 0$ can be taken. Hence, $(\eta^{\Pi(4)}, v^{\Pi(4)})$ constitutes a generalized eigenmode of *matrix A. In fact, since all entries of $\eta^{\Pi(4)}$ are equal, it follows that $A \otimes v^{\Pi(4)} = \frac{1}{2} \otimes v^{\Pi(4)}$, so that $v^{\Pi(4)}$ is an eigenvector of the matrix A for the eigenvalue $\lambda = 5\frac{1}{2}$.*

Example 6.2.2   *Next consider the matrix A studied in Example 5.1.3, given by*

$$A = \begin{pmatrix} 6 & 2 & \varepsilon & 7 \\ \varepsilon & 3 & 5 & \varepsilon \\ \varepsilon & 4 & \varepsilon & 3 \\ \varepsilon & 2 & 8 & \varepsilon \end{pmatrix}.$$

The graph $\mathcal{G}(A)$ is depicted in Figure 5.2, where only the weight of the self-loop at node 1 has to be changed from 1 into 6. Recall that the graph is not strongly connected and that matrix A is reducible.

Take as a starting point policy $\Pi(1) = \{(1,1), (2,2), (2,3), (2,4)\}$ and $v^{\Pi(0)} := \mathbf{u}$. The graph $\mathcal{G}(A^{\Pi(1)})$ associated with $A^{\Pi(1)}$, depicted in Figure 6.5, contains two circuits. One circuit is given by the arc $(1,1)$ and has average weight equal to

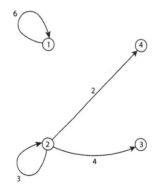

Figure 6.5: Communication graph of $A^{\Pi(1)}$.

6. The other circuit is given by the arc $(2,2)$ and has average weight equal to 3. Further note that nodes 3 and 4 can be reached from node 2. First, consider node 1 and set $\eta_1^{\Pi(1)} := 6$ and $v_1^{\Pi(1)} := 0$. Next, consider node 2 and set $\eta_2^{\Pi(1)} := 3$ and $v_2^{\Pi(1)} := 0$. Now, performing step 4 of Algorithm 6.1.1 starting from node 2 ultimately results in a generalized eigenmode $(\eta^{\Pi(1)}, v^{\Pi(1)})$ of $A^{\Pi(1)}$, with

$$\eta^{\Pi(1)} = \begin{pmatrix} 6 \\ 3 \\ 3 \\ 3 \end{pmatrix} \qquad \text{and} \qquad v^{\Pi(1)} = \begin{pmatrix} 0 \\ 0 \\ 1 \\ -1 \end{pmatrix}.$$

Next, we check whether $(\eta^{\Pi(1)}, v^{\Pi(1)})$ is also an eigenmode of A itself by applying steps 1 and 2 of Algorithm 6.1.2. First, the set I_1 has to be determined for $(\eta^{\Pi(1)}, v^{\Pi(1)})$ as given above. It follows here easily that $I_1 = \emptyset$. Also, it follows that the arc set $\mathcal{D}(\bar{A})$ coincides with $\mathcal{D}(A)$ except for the arcs $(2,1)$ and $(4,1)$ (i.e., $\mathcal{D}(\bar{A}) = \mathcal{D}(A) \backslash \{(2,1) \cup (4,1)\}$). Hence, step 1 can be skipped and step 2 can *be started. So, the set I_2 has to be determined. Therefore, determine $w^{\Pi(1)}$ with components* $w_i^{\Pi(1)} = \max\limits_{(j,i)\in\mathcal{D}(\bar{A})} \{a_{ij} + v_j^{\Pi(1)} - \eta_j^{\Pi(1)}\}$, *for all $i \in \underline{4}$. It follows that*

$$w^{\Pi(1)} = \begin{pmatrix} 0 \\ 3 \\ 1 \\ 6 \end{pmatrix}.$$

Comparing $w^{\Pi(1)}$ with $v^{\Pi(1)}$, as done in step 2, yields $I_2 = \{2, 4\}$. The corresponding arc sets $\mathcal{D}(A)_i^2$ are given by $\mathcal{D}(A)_2^2 = \{(3,2)\}$ and $\mathcal{D}(A)_4^2 = \{(3,4)\}$. Using the improvement as in step 2, it follows that $\Pi(2) = \{(1,1),(3,2),(2,3),(4,3)\}$ has to be taken. The graph $\mathcal{G}(A^{\Pi(2)})$ associated with $A^{\Pi(2)}$ is depicted in Figure 6.6. Repeated application of the steps of Howard's algorithm now yields the

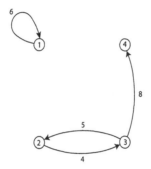

Figure 6.6: Communication graph of $A^{\Pi(2)}$.

following sequence of generalized eigenmodes, communication graphs, and improved policies. The details are left to the reader.
   The generalized eigenmode $(\eta^{\Pi(2)}, v^{\Pi(2)})$ of $A^{\Pi(2)}$ is given by

$$\eta^{\Pi(2)} = \begin{pmatrix} 6 \\ 4\frac{1}{2} \\ 4\frac{1}{2} \\ 4\frac{1}{2} \end{pmatrix} \quad \text{and} \quad v^{\Pi(2)} = \begin{pmatrix} 0 \\ 0 \\ -\frac{1}{2} \\ 3 \end{pmatrix}.$$

The improved policy is given by $\Pi(3) = \{(1,1),(3,2),(4,3),(3,4)\}$. The graph $\mathcal{G}(A^{\Pi(3)})$ associated with $A^{\Pi(3)}$ is depicted in Figure 6.7. The generalized eigenmode $(\eta^{\Pi(3)}, v^{\Pi(3)})$ of $A^{\Pi(3)}$ is given by

$$\eta^{\Pi(3)} = \begin{pmatrix} 6 \\ 5\frac{1}{2} \\ 5\frac{1}{2} \\ 5\frac{1}{2} \end{pmatrix} \quad \text{and} \quad v^{\Pi(3)} = \begin{pmatrix} 0 \\ -1 \\ -\frac{1}{2} \\ 2 \end{pmatrix}.$$

It is now easily checked that the pair $(\eta^{\Pi(3)}, v^{\Pi(3)})$ satisfies (6.9) and (6.10) for all $i \in \underline{4}$, so that

$$A \otimes (v^{\Pi(3)} + k \times \eta^{\Pi(3)}) = v^{\Pi(3)} + (k+1) \times \eta^{\Pi(3)},$$

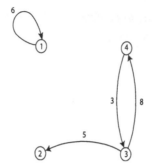

Figure 6.7: Communication graph of $A^{\Pi(3)}$.

for all $k \geq K$, with $K \geq 0$ large enough. By simple verification it follows that here $K = 6$ can be taken. Hence, $(\eta^{\Pi(4)}, v^{\Pi(4)} + 6 \times \eta^{\Pi(4)})$ constitutes a generalized eigenmode of matrix A.

6.3 HOWARD'S ALGORITHM FOR HIGHER-ORDER MODELS

The recurrence relation $x(k + 1) = A \otimes x(k)$ for all $k \geq 0$, with $x(0) = x_0$, can also be formulated as

$$x(k) = A \otimes x(k - 1), \tag{6.12}$$

for all $k \geq 0$, with $x(-1) = x_0$, and Howard's algorithm then yields vectors η and v in $\mathbb{R}^n$ such that for all $k \geq 0$

$$v + k \times \eta = A \otimes (v + (k - 1) \times \eta). \tag{6.13}$$

In this section we are going to study the generalization of (6.12) to higher-order recurrence relations and sketch the development of a Howard-like algorithm for computing a generalized eigenmode. Consider the higher-order recurrence relation

$$x(k) = \bigoplus_{l=0}^{M} A_l \otimes x(k - l), \tag{6.14}$$

for all $k \geq 0$ and with $x(-1), x(-2), \ldots, x(-M)$ specified. For details, see Section 4.5. The matrices $A_0, A_1, \ldots, A_M$ are supposed to belong to the set $\mathbb{R}_{max}^{n \times n}$, where the graph of matrix A_0 does not contain circuits with a nonnegative weight. The latter requirement ensures that the sequence $\{x(k) : k \geq 0\}$ is uniquely determined, given $x(-1), x(-2), \ldots, x(-M)$. The problem of finding the generalized eigenmode then is to find vectors η and v in $\mathbb{R}^n$ such that

$$v + k \times \eta = \bigoplus_{l=0}^{M} A_l \otimes (v + (k - l) \times \eta), \tag{6.15}$$

for all $k \geq 0$. As in Section 6.1.2, the latter equations for η and v can be replaced by two sets of alternative equations. In order to describe these sets we introduce the so-called multigraph associated with (6.14). To abbreviate the notation, we write $\mathcal{A} = (A_0, \ldots, A_M)$ for the matrices in (6.15). The multigraph associated to (6.14) will be denoted by $G(\mathcal{A})$ and consists of a set of nodes $\mathcal{N}(\mathcal{A}) = \underline{n}$ and a set of directed edges $\mathcal{D}(\mathcal{A}) = \{(j, l, i) \in \mathcal{N}(\mathcal{A}) \times \mathbb{N} \times \mathcal{N}(\mathcal{A}) : [A_l]_{ij} \neq \varepsilon\}$. In the latter, (j, l, i) has to be interpreted as an edge from node j to node i with a delay l and a weight $[A_l]_{ij}$. It is therefore possible to have several arcs between node j and i, each with a different delay.

A concise way of writing (6.14) is by means of a shift operator. We formally set $\gamma^{\otimes l} \otimes x(k) = x(k - l)$, $l \geq 0$, and

$$\mathcal{A}(\gamma) = \bigoplus_{l=0}^{M} A_l \otimes \gamma^{\otimes l}.$$

Then, (6.14) reads as $x(k) = \mathcal{A}(\gamma) \otimes x(k)$.

Similar to the analysis put forward in Section 6.1.2, it can be shown that the existence of η and v in $\mathbb{R}^n$ satisfying (6.15) for all $k \geq 0$ is equivalent to the existence of η and v in $\mathbb{R}^n$ such that for all $i \in \underline{n}$

$$\max_{(j,l,i) \in \mathcal{D}(\mathcal{A})} \eta_j = \eta_i, \tag{6.16}$$

$$\max_{(j,l,i) \in \mathcal{D}(\bar{\mathcal{A}})} \left\{ [A_l]_{ij} + v_j - l \times \eta_j \right\} = v_i, \tag{6.17}$$

where $\bar{\mathcal{A}} = (\bar{A}_0, \ldots, \bar{A}_M)$ with $\bar{A}_0, \ldots, \bar{A}_M$ defined by

$$[\bar{A}_l]_{ij} \stackrel{\text{def}}{=} \begin{cases} [A_l]_{ij} & \text{if } (j, l, i) \in \mathcal{D}(\mathcal{A}) \text{ and } \eta_i = \eta_j, \\ \varepsilon & \text{otherwise.} \end{cases}$$

Note that the maximization in (6.16) over arcs in $\mathcal{D}(A)$ that go to node i can equivalently be formulated as a maximization over all nodes j and delays l such that $(j, l, i) \in \mathcal{D}(A)$. A similar remark can be made with respect to (6.17).

In the same spirit as in Section 6.1, a policy improvement algorithm can be developed to solve equations (6.16) and (6.17) for η and v in $\mathbb{R}^n$, where a policy again is a selection of edges such that each node is the end node of precisely one edge. Different edges may share the starting node and may have different delays. Formally, Π is a policy when for each $i \in \mathcal{N}(\mathcal{A})$ there exists an $l \in \mathbb{N}$ and a $j \in \mathcal{N}(\mathcal{A})$ such that $(j, l, i) \in \mathcal{D}(\mathcal{A})$, in which case the edge (j, l, i) is said to belong to Π. We write this as $(j, l, i) \in \Pi$.

Given a policy Π, define the polynomial matrix $\mathcal{A}^\Pi(\gamma) = \bigoplus_{l=0}^{M} A_l^\Pi \otimes \gamma^{\otimes l}$, where

$$[A_l^\Pi]_{ij} \stackrel{\text{def}}{=} \begin{cases} [A_l]_{ij} & \text{if } (j, l, i) \in \Pi, \\ \varepsilon & \text{if } (j, l, i) \notin \Pi. \end{cases}$$

As in Section 6.1.1, it turns out to be easy to compute the generalized eigenmode of $\mathcal{A}^\Pi(\gamma)$; that is, there is a straightforward version of the value determination algorithm, like Algorithm 6.1.1, to compute η and v in $\mathbb{R}^n$ such that for all $k \geq 0$

$$v + k \times \eta = \bigoplus_{l=0}^{M} A_l^\Pi \otimes (v + (k - l) \times \eta).$$

Such η and v satisfy for all $i \in \underline{n}$

$$\max_{(j,l,i) \in \Pi} \eta_j = \eta_i, \qquad \max_{(j,l,i) \in \Pi} \left\{ [A_l]_{ij} + v_j - l \times \eta_j \right\} = v_i.$$

The idea again is check whether the two vectors η and v also satisfy (6.16) and (6.17) for all $i \in \underline{n}$; that is, whether it holds that

$$\max_{(j,l,i) \in \mathcal{D}(\mathcal{A})} \eta_j = \eta_i, \qquad \max_{(j,l,i) \in \mathcal{D}(\bar{A})} \left\{ [A_l]_{ij} + v_j - l \times \eta_j \right\} = v_i,$$

for all $i \in \underline{n}$. If this is the case, a generalized eigenmode has been found. If not, an improved policy is required, as in Section 6.1.2.

The improvement is first based on whether η satisfies (6.16) for all $i \in \underline{n}$. If not, an improved policy can be found as in step 1 of Algorithm 6.1.2, whereupon the value determination algorithm has to be recalled to obtain a new candidate eigenmode and the process has to be repeated.

If η is such that (6.16) is satisfied for all $i \in \underline{n}$, then η and v have to be checked to satisfy (6.17) for all $i \in \underline{n}$. If this is not true, then an improved policy can be found as in step 2 of Algorithm 6.1.2, whereupon the value determination algorithm has to be recalled to obtain a new candidate eigenmode and the process has to be repeated.

6.4 EXERCISES

1. Show that regularity of A implies that the set $\mathcal{D}(\bar{A})$ is not empty.

2. Formulate a version of Howard's algorithm for higher-order descriptions.

3. Develop a version of the power algorithm for higher-order descriptions.

4. Investigate the possibilities to modify Karp's algorithm and the power algorithm so that reducible matrices can be treated. Assume that you have a method to compute the reduced graph and the maximal strongly connected components.

5. Verify that a generalized eigenmode (η, v) of matrix A in Example 2.1.3 is given by

$$\eta = \left(\frac{1}{2}, \frac{1}{2}, \frac{1}{2}, \frac{1}{2}, \frac{4}{3}, \frac{4}{3}, \frac{4}{3}, \frac{4}{3}, \frac{4}{3}, \frac{1}{2} \right)^\top$$

and

$$v = \left(8\frac{1}{2}, 9, 12\frac{1}{2}, 8, 24, 30\frac{1}{3}, 31\frac{2}{3}, 30\frac{5}{6}, 35, 17 \right)^\top.$$

6. Compute a generalized eigenmode (η, v) of the matrix

$$A = \begin{pmatrix} 2 & 1 & 0 & 1 & 1 \\ \varepsilon & 2 & 4 & \varepsilon & \varepsilon \\ \varepsilon & 4 & 1 & \varepsilon & \varepsilon \\ \varepsilon & \varepsilon & \varepsilon & 2 & \varepsilon \\ 5 & 0 & \varepsilon & 0 & 2 \end{pmatrix}.$$

$\left(\text{Answer}: \eta = (4,4,4,2,4)^\top \text{ and } v = (-3,0,0,0,-2)^\top. \right)$

7. Compute a generalized eigenmode (η, v) for the higher-order recurrence corresponding to equation (0.13);

$$x(k) = \begin{pmatrix} 2 & 5 \\ \varepsilon & 3 \end{pmatrix} \otimes x(k-1) \oplus \begin{pmatrix} \varepsilon & \varepsilon \\ 3 & \varepsilon \end{pmatrix} \otimes x(k-2), \qquad k \geq 2.$$

$\left(\text{Answer}: \eta = (3,3)^{\top} \text{ and } v = (2,0)^{\top}.\right)$

8. Compute a generalized eigenmode (η, v) for the higher-order recurrence

$$x(k) = \begin{pmatrix} \varepsilon & 1 \\ -4 & \varepsilon \end{pmatrix} \otimes x(k) \oplus \begin{pmatrix} 2 & 4 \\ -1 & \varepsilon \end{pmatrix} \otimes x(k-2), \qquad k \geq 2.$$

Do the computations by applying the ideas in Section 6.3. Verify the results by going over to a first-order recurrence relation and by applying Howard's algorithm as presented in Section 6.1. $\left(\text{Answer}: \eta = (1,1)^{\top} \text{ and } v = (0,-3)^{\top}.\right)$

6.5 NOTES

The present chapter is based upon [23]. There, details can also be found on the convergence of Howard's algorithm. However, no tight bounds can be given for the complexity of the algorithm. Numerical experiments indicate that Howard's algorithm is extremely efficient and should be preferred to other algorithms. We also refer to [23] for details on Howard's algorithm for higher-order descriptions. Howard's algorithm is named for the author of [54]. See [33] for a convergence proof.

All three authors of this book were involved in a European project called Alapedes (the ALgebraic Approach to Performance Evaluation of Discrete Event Systems). At one of the Alapedes meetings, in Waterford, Ireland, in 1997, another Alapedes member, Stéphane Gaubert, showed for the first time the power of Howard's algorithm on his laptop. The program calculated in virtually no time the eigenmodes of very large matrices (number of rows and columns in the thousands). We all were greatly impressed. See www.maxplus.org under scilab for information on the software concerned.

PART II
Tools and Applications

Chapter Seven

Petri Nets

In this chapter we will give a brief introduction to Petri nets as a modeling tool. We will show that a subclass of Petri nets, the so-called event graphs, is a suitable modeling aid for the construction of max-plus linear systems (i.e., for the construction of equations like (0.9) or (4.7)). In Section 7.1, the definitions of a Petri net and a timed event graph will be given. The construction of max-plus linear systems, starting from an event graph description of a model, will be treated in Section 7.2 for the autonomous case (i.e., when no external inputs are considered), and in Section 7.3 for the nonautonomous case.

7.1 PETRI NETS AND EVENT GRAPHS

Max-plus algebra allows us to describe the evolution of events on a network subject to synchronization constraints. For a railway network, for instance, the departures of trains are events. An appropriate tool to model events on a certain class of networks is named after C. A. Petri. This class of networks, to be introduced shortly, is therefore called the class of *Petri nets*. A subclass of these Petri nets, called *event graphs*, can be modeled by max-plus linear recurrence relations. Specifically, an event graph description can be transformed into a max-plus model and vice versa.

Petri nets are special directed graphs. The (finite) set of nodes $\mathcal{N}$ can be partitioned into two disjoint subsets $\mathcal{P}$ and $\mathcal{Q}$. The elements of $\mathcal{P}$ are called places and those of $\mathcal{Q}$ are called transitions. Places will be denoted by p_i, $i = 1, 2, \ldots, |\mathcal{P}|$, and transitions will be denoted by q_j, $j = 1, 2, \ldots, |\mathcal{Q}|$. Arcs from places to transitions exist $(p_i \rightarrow q_j)$, as well as from transitions to places $(q_i \rightarrow p_j)$, but arcs from places to places or from transitions to transitions do not exist. Hence, the set of arcs, to be denoted by $\mathcal{D}$, satisfies $\mathcal{D} \subset (\mathcal{Q} \times \mathcal{P}) \cup (\mathcal{P} \times \mathcal{Q})$. The set $\mathcal{Q} \times \mathcal{P}$ consists of all elements (q_i, p_j); the set $\mathcal{P} \times \mathcal{Q}$ is likewise defined. Because of this latter property, one says that Petri nets are bipartite directed graphs.

For $(p_i, q_j) \in \mathcal{D}$ we say that node p_i is an upstream place for q_j, and q_j is a downstream transition for p_i. The nodes of $(q_j, p_i) \in \mathcal{D}$ are similarly expressed. In agreement with the notation introduced in Chapter 2, we denote the set of all upstream places of transition q_j by $\pi(q_j)$, i.e., $p_i \in \pi(q_j)$ if and only if $(p_i, q_j) \in \mathcal{D}$. Similarly, the set of all upstream transitions of place p_i is denoted by $\pi(p_i)$; i.e., $q_j \in \pi(p_i)$ if and only if $(q_j, p_i) \in \mathcal{D}$. Downstream relationships are defined analogously by means of the symbol σ instead of π. See also Section 2.1, where these symbols were introduced. A Petri net is called an event graph if all places in the Petri net have exactly one upstream and one downstream transition. Transitions

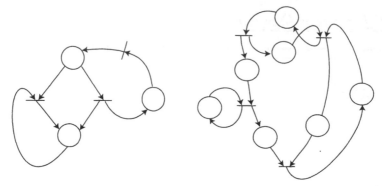

Figure 7.1: A Petri net (left) and an event graph (right). The circles represent places and the bars transitions.

in an event graph can have more than one upstream and/or downstream place. See Figure 7.1 for an example of a Petri net and of an event graph. In such graphical representations, places are always drawn as circles and transitions as bars. These bars can have any orientation (not only horizontal or vertical).

In terms of applications, places often represent conditions and transitions represent events. A condition being fulfilled is indicated by a token allocated to the corresponding place (in the graph this is indicated by a dot in the circle representing the place concerned). The event symbolized by a transition can take place (in Petri net terminology, we say that the transition is *enabled*) if all upstream places contain at least one token. If a transition is enabled, then it can execute the event it symbolizes. If the event takes place, one says that the transition *fires*. If an event happens (i.e., the corresponding transition fires), then a token is taken away from each of the upstream places and one token is added to each of the downstream places. An example is given in Figure 7.2. Note that if the number of upstream places differs from the number of downstream places of a transition, the total number of tokens before and after the firing will be different. The number of tokens in each of the places, called the marking of the Petri net, is indicated by the vector $\mathcal{M} = (m_1, m_2, \ldots, m_{|\mathcal{P}|})^\top$. Each element m_i is a natural number. If a transition fires, the vector $\mathcal{M}$ will in general change. The initial marking is indicated by $\mathcal{M}_0$. Usually the order of firings is not uniquely determined.

A Petri net (or event graph) is called a *timed Petri net* (or *timed event graph*) if a holding time is attached to each place. This holding time associated with place p_i will be indicated by τ_i. It represents the time that a token must spend in the place before it can play its role in the enabling of the downstream transition. (If there is more than one place upstream of this transition, then the tokens in all these places each must have spent their holding times before the transition is enabled.) This firing, or event happening, is supposed to be instantaneous, i.e., it does not take time. We also say that the firing time equals zero. To be very explicit, time durations are only attached to places and not to transitions. The vector whose elements are the holding times will be indicated by $\mathcal{T}$.

Summing up, we get the following definition.

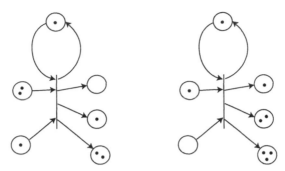

Figure 7.2: The token distribution before (left) and after (right) a firing of a transition. Note that if a place is both up- and downstream, the number of tokens before and after the firing will be the same.

DEFINITION 7.1 *A timed Petri net* $\mathcal{G}$ *is characterized by* $\mathcal{P}, \mathcal{Q}, \mathcal{D}, \mathcal{M}_0,$ *and* $\mathcal{T},$ *where* $\mathcal{P}$ *is the set of places,* $\mathcal{Q}$ *is the set of transitions,* $\mathcal{D}$ *is the set of arcs from transitions to places and vice versa,* $\mathcal{M}_0$ *is the initial marking, and* $\mathcal{T}$ *is the vector of holding times. If each place has exactly one upstream and one downstream transition, then the (timed) Petri net is called a (timed) event graph.*

From this point on, we will restrict ourselves to the study of event graphs. In event graphs, the location of each place is characterized by its upstream and downstream transitions. If only one place between two transitions exists (keeping track of the directions of the arcs), it is uniquely determined by these transitions. For this reason a place p_l will also be indicated by the notation p_{ji}, where the subscript i refers to the upstream transition q_i and j to the downstream transition q_j. The same convention of notation will be used for holding times and markings. Only in the case where there is more than one place between two transitions are the notations $p_{ji}, \tau_{ji},$ and m_{ji} ambiguous.

THEOREM 7.2 *The number of tokens in any circuit of an event graph is constant.*

Proof. This is straightforward. $\Box$

As an immediate consequence of the above theorem, if a circuit contains zero tokens, then the transitions within this circuit will never fire. An event graph is called *live* if each circuit contains at least one token. See exercise 3 at the end of the chapter for liveness of Petri nets.

Example 7.1.1 *In Figure 7.3, Figure 0.1 is repeated in more abstract form (left), and the corresponding event graph is given on the right. The stations (or more properly, the departures of the trains per track) are indicated by transitions. A train on a track is symbolized by a token in the place on this track. Saying that the trains at station* S_1 *are leaving is equivalent to saying that transition* q_1 *fires. In order for it to fire, the incoming trains must have arrived or, equivalently, the tokens in the upstream places must have spent their holding time. The holding time is here defined to be the travel time (in which the changeover time has been subsumed).*

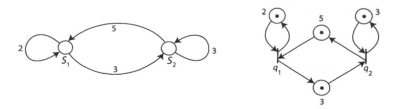

Figure 7.3: Figure 0.1 (left) and the corresponding timed event graph (right).

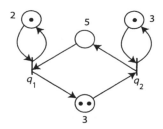

Figure 7.4: Figure 7.3 (right) after one firing.

Once transition q_1 has fired, the token distribution becomes as given in Figure 7.4. It will be obvious that transition q_1 cannot immediately fire a second time, since the place between transition q_2 (station S_2) and transition q_1 no longer contains a token. In order for transition q_1 to fire a second time, this place must have a token, which will be the case after transition q_2 has fired (and the holding time has been spent).

Example 7.1.2 *In Figure 7.5, the event graphs corresponding to (0.13), (0.15), (0.16), and exercise 6 of Chapter 0 are given. Let us consider the construction of the graphs corresponding to (0.13) and (0.16) in some detail. The other two are left as exercises.*

Equation (0.13) represents a two-dimensional system; hence, we have two transitions, one corresponding to each of the two states. Because there are four connections (line segments), the event graph will contain four places. All places contain one token (interpret a token as a train), except for the place between transition q_1 and transition q_2, which has two tokens. This is due to the argument $k - 1$ of $x_1(k - 1)$ at the right-hand side of the second equation of (0.13). The numbers attached to the places in the figure refer to the travel times.

Equation (0.16) represents a three-dimensional system, and hence, there are three transitions. Places exist between those transitions for which the corresponding element in the system matrix in (0.16) is finite. All these places contain exactly one token, which is due to the fact that (0.16) is a first-order system.

Note that the definition of a timed event graph as given does not uniquely determine all (future) firing times. In order to achieve that, it is necessary to add initial

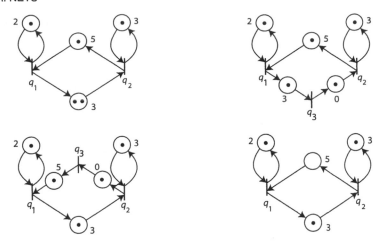

Figure 7.5: The event graphs corresponding to (0.13), (0.15), (0.16), and exercise 8 of Chapter 0.

conditions; to be precise, for every token one should specify in which place it finds itself and how long it has already been there (that is, how much of its holding time it has already consumed).

In the examples considered so far, all transitions had upstream and downstream places. It is quite possible to study event graphs with transitions that do not have such places. A transition without an upstream place is called a *source transition*, or simply a *source*. Such a transition is supposed to be enabled by the outside world. Similarly, a transition that does not have a downstream place is called a *sink transition*) (or a *sink*). Sink transitions deliver tokens to the outside world. If there are no sources in the network, then one talks about an *autonomous* network and calls it *nonautonomous* otherwise. It is assumed that only transitions can be sources or sinks (which is no loss of generality, since one can always add a transition upstream or downstream of a place if necessary). A source transition is an input of the network; a sink transition is an output of the network.

7.2 THE AUTONOMOUS CASE

As before, let τ_{ji} denote the holding time of the place between transition q_i and q_j, tacitly assuming that there is only one such place, and define $a_{ji} \overset{\text{def}}{=} \tau_{ji}$. Though generalizations are possible, here the focus is on constant holding times, i.e., they are independent of the counter k. Let $x_j(k)$ denote the time at which transition q_j fires for the kth time. We can define the vector $x(k) = (x_1(k), \ldots, x_{|\mathcal{Q}|}(k))^\top$ as the state of the system.

With any event graph, matrices $A_0, \ldots, A_M$ can be associated, all of size $|\mathcal{Q}| \times |\mathcal{Q}|$. To obtain $[A_m]_{jl}$ consider all places between transitions q_l and q_j having initially m tokens, and take as $[A_m]_{jl}$ the maximum of the corresponding holding times. It is possible that more than one place between two transitions exists. In such a case, $[A_m]_{jl} = a_{jl}$, where a_{jl} is the largest of the holding times with respect to all places between transitions q_l and q_j with m tokens. In other words, when there is only one such a place,

$$[A_m]_{jl} = \begin{cases} a_{jl} & \text{if the number of tokens in place } p_{q_j q_l} \text{ equals } m, \\ \varepsilon & \text{otherwise,} \end{cases} \tag{7.1}$$

for $m = 0, 1, \ldots, M$, where M is defined as the maximum number of tokens with respect to all places.

If one considers the state variable $x_i(k)$, which denotes the kth time that transition q_i fires, then the vector $x(k) = (x_1(k), \ldots, x_{|\mathcal{Q}|}(k))^\top$ satisfies the max-plus (linear) equation

$$x(k) = A_0 \otimes x(k) \oplus A_1 \otimes x(k-1) \oplus \cdots \oplus A_M \otimes x(k-M), \tag{7.2}$$

for $k \geq 0$.

Remark. The matrices A_m in (7.2) were defined with respect to a token distribution at the initial time (or at another time instant). To start with, one could take a picture of the whole net and count the number of tokens in each of the places in this picture and subsequently construct the A_m matrices. When k increases by one, each transition has fired once, and the token distribution will again be the same. This is not to be confused with the firing of a single transition. Compare, for instance, the token distributions of Figures 7.3 (right) and 7.4. The system descriptions based on these two figures will be different, though the same instants at which tokens move is described. This feature was already encountered in Section 0.4. In conventional algebra one has a somewhat similar phenomenon: a coordinate transformation yields different mathematical representations of the same system.

If we follow the train of thought put forward in Section 4.5, then equation (7.2) can be written as

$$x(k) = A_0^* \otimes A_1 \otimes x(k-1) \oplus \cdots \oplus A_0^* \otimes A_M \otimes x(k-M), \tag{7.3}$$

for $k \geq 0$, where

$$A_0^* = \bigoplus_{i=0}^{|\mathcal{Q}|-1} A_0^i, \tag{7.4}$$

provided that the communication graph of A_0 does not contain circuits with positive weight.

As a next step, transform (7.3) into a first-order recurrence relation. In order to do so, we take as new state vector the $(|\mathcal{Q}| \times M)$–dimensional vector

$$\tilde{x}(k) = (x^\top(k), x^\top(k-1), \ldots, x^\top(k-M+1))^\top \tag{7.5}$$

and the $(|\mathcal{Q}| \times M) \times (|\mathcal{Q}| \times M)$–dimensional matrix $\tilde{A}$ given by

$$\begin{pmatrix} A_0^* \otimes A_1 & A_0^* \otimes A_2 & \cdots & \cdots & A_0^* \otimes A_M \\ E & \varepsilon & \cdots & \cdots & \varepsilon \\ \varepsilon & E & \varepsilon & \cdots & \varepsilon \\ \vdots & \ddots & \ddots & \ddots & \vdots \\ \vdots & & & & \vdots \\ \varepsilon & \cdots & \varepsilon & E & \varepsilon \end{pmatrix}.$$

Then (7.3) can be written as

$$\tilde{x}(k) = \tilde{A} \otimes \tilde{x}(k-1), \qquad k \geq 0. \tag{7.6}$$

The above equation is called the *standard autonomous equation*. Any live autonomous event graph can be modeled by a standard autonomous equation.

Remark. Equation (7.3) is a recurrence relation (by abuse of language it is often called a *difference equation*). If numerical values for the vectors $x(-1)$, $x(-2), \ldots, x(-M)$ are given, then these values constitute the initial condition, and the future evolution of the state is uniquely determined. In the same way, $\tilde{x}(-1)$ is an initial condition for (7.6). From a mathematical point of view, there are no restrictions on the numerical values for these initial conditions. Given a physical interpretation, however, limitations for these values may exist. An example is the vector of initial conditions of which the components represent both arrival and departure times. The arrival time of a train at a station should occur before the departure time of the same train.

Sometimes the token distribution at a certain time instant is interpreted as an initial condition (compare taking a picture of the whole net in the previous remark). This is not fully correct, however, since one does not know how long a token has already resided in its place with only this information. In other words, one does not know how much of the holding time has already been consumed. One could say that the token distribution only provides partial information for the initial condition.

Often the dimension of the obtained state space is unnecessarily large. Usually quite a few places in the original event graph can be combined into a single place. As an example, two places in a series with a transition in between that has neither upstream nor downstream places can be replaced by a single place. This causes a smaller $|\mathcal{Q}|$ and, hence, a smaller dimension. Algorithm 8.3.1 in Chapter 8 describes this reduction of the dimension of the state space in a systematic way.

Example 7.2.1 *Let us consider a circular track with three stations along which two trains run in one direction. The trains run from station S_1 to S_2, from S_2 to S_3, from S_3 to S_1, and so on. For safety reasons it is assumed that a train cannot leave station S_i before the preceding train has left S_{i+1}, $i \in \underline{3}$, with $S_4 = S_1$, or in other words, a train cannot leave a station before the platform at the next station is free. This model is symbolized in the Petri net in Figure 7.6, in which the transitions are denoted by S_i, $i \in \underline{3}$. The trains move counterclockwise, whereas the tokens in the places in the clockwise circuit represent the conditions of the next station being free. If $x_i(k)$ represents the kth departure from station S_i and $a_{i+1,i}$, $i \in \underline{3}$, with $a_{4,3} = a_{1,3}$, is the travel time between S_i and S_{i+1}, then*

$$x_1(k+1) = \max\{x_3(k+1) + a_{13}, x_2(k+1)\},$$
$$x_2(k+1) = \max\{x_1(k) + a_{21}, x_3(k+1)\},$$
$$x_3(k+1) = \max\{x_2(k) + a_{32}, x_1(k)\},$$

which, like (7.2) and (7.4) can be written as

$$x(k+1) = \begin{pmatrix} \varepsilon & 0 & a_{13} \\ \varepsilon & \varepsilon & 0 \\ \varepsilon & \varepsilon & \varepsilon \end{pmatrix} \otimes x(k+1) \oplus \begin{pmatrix} \varepsilon & \varepsilon & \varepsilon \\ a_{21} & \varepsilon & \varepsilon \\ 0 & a_{32} & \varepsilon \end{pmatrix} \otimes x(k)$$

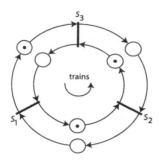

Figure 7.6: A Petri net of a circular track with two trains.

$$= \begin{pmatrix} a_{21} \oplus a_{13} & a_{13} \otimes a_{32} & \varepsilon \\ a_{21} & a_{32} & \varepsilon \\ 0 & a_{32} & \varepsilon \end{pmatrix} \otimes x(k). \tag{7.7}$$

In the derivation of the latter equality in (7.7), it has been tacitly assumed that all a_{ij} are nonnegative. Since the evolution of x_1 and x_2 is not influenced by the evolution of x_3 (the third column of the latter system matrix only contains ε's), the reduced state $x_{\mathrm{red}} \stackrel{\mathrm{def}}{=} (x_1, x_2)^\top$ can be introduced, which satisfies

$$x_{\mathrm{red}}(k+1) = \begin{pmatrix} a_{21} \oplus a_{13} & a_{13} \otimes a_{32} \\ a_{21} & a_{32} \end{pmatrix} \otimes x_{\mathrm{red}}(k). \tag{7.8}$$

7.3 THE NONAUTONOMOUS CASE

Let $\mathcal{Q}_I \subset \mathcal{Q}$ denote the set of input transitions, while $\mathcal{Q}$ remains the set of all transitions. If q_i is such an input transition, then the holding time of a place between q_i and another transition $q_j \in \mathcal{Q} \setminus \mathcal{Q}_I$ is indicated by b_{ji}. If the maximal initial marking of all places downstream of any input transition is denoted by M', then $|\mathcal{Q} \setminus \mathcal{Q}_I| \times |\mathcal{Q}_I|$-dimensional matrices $B_0, \ldots, B_{M'}$ are defined as

$$[B_m]_{jl} = \begin{cases} b_{jl} & \text{if the number of tokens in place } p_{q_j q_l} \text{ equals } m, \\ \varepsilon & \text{otherwise,} \end{cases}$$

for $m = 0, 1, \ldots, M'$. In words, to obtain $[B_m]_{jl}$ we consider all upstream transitions of place j that are input transitions having initially m tokens in their preceding place. If there happens to be more than one such place, we define $[B_m]_{jl}$ to be the maximum of the corresponding holding times. Furthermore, we let $u(k)$ be a $|\mathcal{Q}_I|$-dimensional vector, where $u_i(k)$ denotes the kth firing time of the appropriate input transition.

The vector of the kth firing times satisfies the (linear) equation

$$x(k) = A_0 \otimes x(k) \oplus A_1 \otimes x(k-1) \oplus \cdots \oplus A_M \otimes x(k-M)$$
$$\oplus B_0 \otimes u(k) \oplus B_1 \otimes u(k-1) \oplus \cdots \oplus B_{M'} \otimes u(k-M'), \tag{7.9}$$

for $k \geq 0$. A possible initial condition is that $x_j(k)$, $j \in \underline{|\mathcal{Q}| \setminus |\mathcal{Q}_I|}$, and $u_j(k)$, $j \in \underline{|\mathcal{Q}_I|}$, equal ε if $k < 0$.

From now on we restrict ourselves to nonautonomous event graphs for which
the communication graph of A_0 does not contain any circuit with positive weight.
Then (7.9) is equivalent to

$$x(k) = A_0^* \otimes A_1 \otimes x(k-1) \oplus \cdots \oplus A_0^* \otimes A_M \otimes x(k-M)$$
$$\oplus A_0^* \otimes B_0 \otimes u(k) \oplus \cdots \oplus A_0^* \otimes B_{M'} \otimes u(k-M'), \qquad (7.10)$$

for all $k \geq 0$. Compare this recurrence relation with (7.3) for the autonomous case,
and note that now terms with the input firings have been added.

Define the $((M'+1) \times |\mathcal{Q}_I|)$–dimensional vector

$$\tilde{u}(k) = (u^\top(k), u^\top(k-1), \ldots, u^\top(k-M'))^\top$$

and the $(|\mathcal{Q} \setminus \mathcal{Q}_I| \times M) \times (|\mathcal{Q}_I| \times (M'+1))$ matrix $\tilde{B}$ by

$$\begin{pmatrix} A_0^* \otimes B_0 & A_0^* \otimes B_1 & \cdots & \cdots & A_0^* \otimes B_{M'} \\ \mathcal{E} & \mathcal{E} & \cdots & \cdots & \mathcal{E} \\ \vdots & \vdots & \cdots & \cdots & \vdots \\ \mathcal{E} & \mathcal{E} & \cdots & \cdots & \mathcal{E} \end{pmatrix}.$$

Then (7.10) can be written as

$$\tilde{x}(k) = \tilde{A} \otimes \tilde{x}(k-1) \oplus \tilde{B} \otimes \tilde{u}(k), \qquad k \geq 0,$$

where $\tilde{x}$ and $\tilde{A}$ are defined as in the previous section. We call the above equation
the *standard nonautonomous equation*. Any nonautonomous event graph (with the
communication graph of A_0 not having circuits with positive weight) can be modeled by a standard nonautonomous equation.

Example 7.3.1 *Consider a railway system with external arrivals modeling a
single long-distance track with two stations. Let station S_0 represent an external
source generating arrivals of trains, and denote the interarrival time of trains by
a_0. Assume that the distances between stations S_1 and S_2 and between the source
stations S_0 and S_1 are long, so that more than one (actually, arbitrarily many)
trains can be present on the intermediate tracks. The travel times are a_{10} and a_{21},
respectively. Each train that enters the system has to pass through stations S_1 and
S_2, each of which can handle only one train at a time, and then leaves the line.
This is modeled by $a_{11} = a_{22} = e$. We assume that the system starts empty. This
description is symbolized as a Petri net model in Figure 7.7. Let $x_j(k)$ denote the*

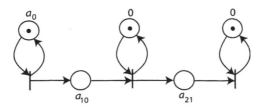

Figure 7.7: A Petri net model of Example 7.3.1.

kth departure time of a train from station S_j, $j \in \underline{2}$. Let $x_0(k)$ denote the time of the kth arrival of a train on the line. The time evolution of the system can then be described by a three-dimensional vector $x(k) = (x_0(k), x_1(k), x_2(k))^\top$ following the homogeneous equation

$$x(k+1) = A \otimes x(k),$$

with $x(0) = \mathbf{u}$ and where

$$A = \begin{pmatrix} a_0 & \varepsilon & \varepsilon \\ a_0 \otimes a_{10} & e & \varepsilon \\ a_0 \otimes a_{10} \otimes a_{21} & a_{21} & e \end{pmatrix}.$$

Observe that A is reducible. Alternatively, we could describe the system via a two-dimensional vector $x_{\mathrm{red}}(k)$ following the inhomogeneous equation

$$x_{\mathrm{red}}(k+1) = \begin{pmatrix} e & \varepsilon \\ a_{21} & e \end{pmatrix} \otimes x_{\mathrm{red}}(k) \oplus \begin{pmatrix} a_{10} \\ a_{10} \otimes a_{21} \end{pmatrix} \otimes u(k+1),$$

for $k \geq 0$, where $u(k) = a_0^{\otimes k} = k \times a_0$ denotes the time of the kth arrival of a train on the line.

7.4 EXERCISES

1. Give the Petri net representation of (0.14).

2. Give a proof of Theorem 7.2.

3. A Petri net is said to be *live* (for the initial marking $\mathcal{M}_0$) if for any $\mathcal{M}$, obtained after an arbitrary series of firings starting from $\mathcal{M}_0$, and for each transition q there exists another marking $\mathcal{N}$, which can be obtained after a suitable series of firings starting from $\mathcal{M}$, such that q is enabled in $\mathcal{N}$. Prove the following:

   - For a live Petri net, or a live event graph, any transition can be fired an infinite number of times.

   - An event graph is live if and only if each circuit contains at least one token.

   - An event graph of an autonomous system is live if and only if the communication graph of A_0 does not have circuits.

4. Calculate the matrices A_m as given in equation (7.2) for both Figures 7.3 (right) and 7.4. Rewrite the model for Figure 7.4 in the format of the standard autonomous equation. Thus, one has two first-order descriptions of the same event graph. Compare the behaviors of these two descriptions.

5. If the travel times in Example 7.2.1 of the kth departure depend on k, i.e., $a_{ij}(k)$, then prove that (7.8) becomes

$$x_{\mathrm{red}}(k+1) = \begin{pmatrix} a_{21}(k) \oplus a_{13}(k+1) & a_{13}(k+1) \otimes a_{32}(k) \\ a_{21}(k) & a_{32}(k) \end{pmatrix} \otimes x_{\mathrm{red}}(k).$$

7.5 NOTES

Petri nets were the subject of the PhD thesis [78] of C. A. Petri. An excellent overview of the theory can be found in [67]. The chapter in this book on Petri nets is only a minimal introduction to the subject. Many books on the subject have been written. A recent one is [32].

Event graphs are sometimes also referred to as *marked graphs* or *decision-free Petri nets*. Typical examples of applications are the G/G/1 queues, networks of (finite) queues in tandem, kanban systems, flexible manufacturing systems, fork/join queues, or any parallel and/or series composition made by these elements.

More abstract system descriptions, with two-domain formal power series, can be found in [26]. As an example, this description for the Petri net of Figure 7.4 is

$$
\begin{pmatrix} x_1 \\ x_2 \end{pmatrix} = \begin{pmatrix} \delta^2\gamma & \delta^5 \\ \delta^3\gamma^2 & \delta^3\gamma \end{pmatrix} \begin{pmatrix} x_1 \\ x_2 \end{pmatrix},
$$

where the $\otimes$ symbol in the exponents is suppressed for ease of notation. The exponents of δ refer to the holding times, and the exponents of γ refer to the number of tokens in the corresponding places; for a formal definition see Section 6.3. The operator γ is sometimes called the *shift operator*. The states x_i are the formal power series in γ and δ. With the rules

$$
t\gamma^k \oplus \tau\gamma^k = \max(t,\tau)\gamma^k, \qquad k\delta^t \oplus k\delta^\tau = k\delta^{\max(t,\tau)},
$$

it is easy to see that the series

$$
\begin{pmatrix} x_1 \\ x_2 \end{pmatrix} = \bigoplus_{k=-\infty}^{+\infty} (\gamma\delta^4)^k \begin{pmatrix} \delta \\ \gamma \end{pmatrix}
$$

is a solution to the (implicit) system description given above.

In Chapter 1, a reference was made to system-theoretical concepts. In systems theory one considers input/output relations. In terms of Petri nets, inputs coincide with source transitions and outputs with sink transitions. Many relationships with (conventional) systems theory exist. Well-known concepts in this theory, on stability, feedback, model reference control, and so on, have found corresponding notions within a max-plus setting; see, [29] for such an example, and www.istia.univ-angers.fr/~hardouin/outils.html for free downloadable software.

Chapter Eight

The Dutch Railway System Captured in a
Max-Plus Model

This chapter and the next deal with the application of max-plus algebra in a study of the timetable of the Dutch railway system. The starting point is the railway track layout, consisting of a number of lines along which trains run up and down, and the requested synchronization data (i.e., which trains should wait for which other trains in order to allow passengers to transfer from one to the other). It will be assumed that this data is provided. In addition, it is assumed that a timetable is given with a period of one hour (or, a frequency of one train per hour). This is somewhat in contrast to the approach in Chapter 0, where the problem of timetable design was the central issue once the railway network and the number of trains on the network were given. In the current chapter, the number of trains has become a variable to be found. Indeed, one of the problems studied is to find the minimum number of trains required to realize the operability of a given timetable. The construction of the appropriate max-plus model is the subject of the current chapter. In the next chapter, various results will be given, and we will deal with some notions of stability or robustness in case of disturbances.

In Section 8.1 the Dutch railway system is introduced and a characterization in terms of line and synchronization data is given. By means of a subnetwork–the whole network is simply too complex to be presented in detail–the transformation of this data into a max-plus model will be elucidated in the coming sections (results for the whole network will be given in Section 9.2). In Section 8.2 the timed event graph will be constructed from the data given, and subsequently the max-plus model will be constructed in Section 8.3. Finally, in Section 8.4 some numerical results will be given.

8.1 THE LINE SYSTEM

The max-plus algebra approach has been applied successfully to the study of timetables of the Dutch railway system. In order to show this, a subnetwork of the Netherlands will be used throughout this chapter. The complete network of the Dutch railway layout for the so-called intercity trains is given in Figure 8.1. The study of this network requires a max-plus model of about one hundred variables, which is not suitable for presentation in a book like this. (If one were also to include the express trains and local trains, a model of several hundreds of variables would result.) Therefore, a subnetwork, presented in Figure 8.2, has been chosen, which

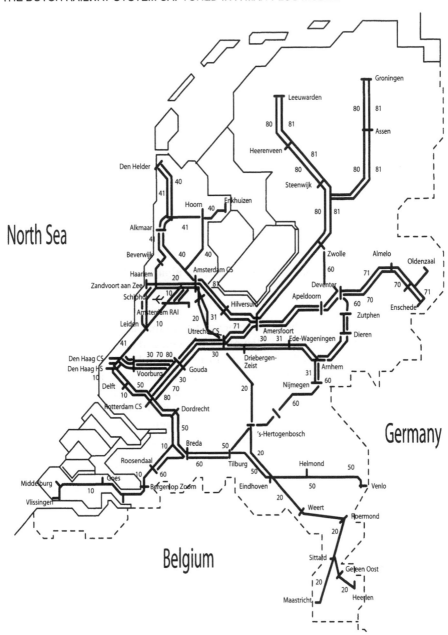

Figure 8.1: The intercity network of the Netherlands. The numbers along the tracks refer to line numbers.

is easier to comprehend in the analysis to come.

Many readers may not be familiar with the geography of the Netherlands. Consider Figure 8.1. As with most maps, it is oriented with north at the top, the east

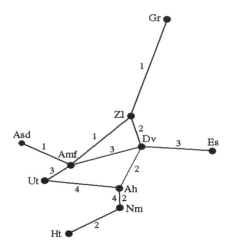

Figure 8.2: The railway network. The numbers along the tracks refer to the line number.

to the rigth, and so on. In the map, Germany lies to the east, Belgium to the south and France is situated south of Belgium. To the west and north, we have the North Sea. To give you an idea of the scale, the distance from Delft (near Rotterdam), in the west, to Groningen, in the north, is about 200 kilometers as the crow flies.

Let us now study Figure 8.2 in some detail. It consists of four long-distance lines on which intercity (IC) trains run with an interval of one hour. The first line goes from Amsterdam Central Station (Asd) to Groningen (Gr) and back. The second line goes from Zwolle (Zl) to 's Hertogenbosch (Ht) and back. The third line goes from Utrecht Central Station (Ut) to Enschede (Es) and back. The last line goes from Utrecht Central Station to Nijmegen (Nm) and back. The numbers of these lines do not coincide with the numbers given in Figure 8.1. The characteristics of this network are provided by the so-called line data, given in Table 8.1, and synchronization data, given in Table 8.2. The algorithms to be described in this chapter assume the availability (by the railway authorities) of two input data files, consisting of line data and synchronization data, respectively, and they will be described now.

Each row of the line data file (see Table 8.1 for an example) represents a particular line segment. The file contains eight columns, which represent the following:

1. the row number in the table,

2. the line number,

3. the segment number of the line,

4. the origin station of the line segment,

5. the destination station of the line segment,

No.	Line	Segm.	Orig.	Dest.	Run	Dwell	Depart.
1	1	01	Asd	Amf	32	2	34
2	1	02	Amf	Zl	36	2	10
3	1	03	Zl	Gr	65	5	49
4	1	51	Gr	Zl	66	2	38
5	1	52	Zl	Amf	36	2	48
6	1	53	Amf	Asd	31	5	28
7	2	01	Zl	Dv	19	2	23
8	2	02	Dv	Ah	35	2	44
9	2	03	Ah	Nm	15	2	21
10	2	04	Nm	Ht	29	5	38
11	2	51	Ht	Nm	28	2	26
12	2	52	Nm	Ah	14	2	56
13	2	53	Ah	Dv	35	2	12
14	2	54	Dv	Zl	20	5	49
15	3	01	Ut	Amf	14	2	52
16	3	02	Amf	Dv	37	2	10
17	3	03	Dv	Es	43	5	49
18	3	51	Es	Dv	43	2	58
19	3	52	Dv	Amf	40	2	44
20	3	53	Amf	Ut	15	5	27
21	4	01	Ut	Ah	34	2	20
22	4	02	Ah	Nm	12	5	59
23	4	51	Nm	Ah	13	2	21
24	4	52	Ah	Ut	34	5	39

Table 8.1: Line data.

6. the run time on the line segment,

7. the waiting time (also called *dwell time*) at the destination station of the line segment, and

8. the departure time (in terms of minutes after the full hour) according to the timetable at the origin station.

The input file also implicitly contains the buffer times, which can be obtained by calculating the arrival time at the destination station and then subtracting this and the dwell time from the subsequent departure time. All times are expressed in minutes.

As an example, a train on line 1 departs from Amsterdam Central Station (Asd) every 34 minutes past the hour heading toward Amersfoort (Amf). The line segment has number 01. The train arrives at Amersfoort 32 minutes later, at 6 minutes past the hour, where it stops for at least 2 minutes (waiting time). The train departs at 10 minutes past the hour on the following line segment 02. Hence, there is a buffer

No.	Feeder Line	Feeder Segm.	Connecting Line	Connecting Segm.	Transfer time
1	1	01	3	02	2
2	1	52	3	53	2
3	2	01	3	03	2
4	3	01	1	02	2
5	3	51	2	54	2
6	3	51	2	02	2
7	3	52	1	53	2

Table 8.2: Synchronization data.

time of 2 minutes that can be used to reduce or even to eliminate a possible arrival delay.

Each row of the synchronization file (see Table 8.2 for an example), consisting of six items, contains a synchronization constraint between two trains. These items are the following:

1. the row number in the table,

2. the line number of the incoming (feeder) train,

3. the segment number of the incoming train,

4. the line number of the leaving (connecting) train,

5. the segment number of the leaving train, and

6. the minimum passenger transfer time required between the arrival of the feeder train and the departure of the connecting train.

As an example let us read the first row. The arriving train on line 1, segment 01, has a transfer connection with the train departing on line 3, segment 02, with a transfer time of 2 minutes. From the line data file, we can identify the line segments and calculate the arrival time of the arriving train and the departure time of the departing train. The feeder train on line 1 comes from Amsterdam Central Station and arrives at Amersfoort at $(34 + 32) \bmod 60 \equiv 6$ minutes past every hour (cf. row 1 in Table 8.1). The connecting train on line 3 departs at Amersfoort heading towards Deventer (Dv) every 10 minutes past the hour (cf. row 16 in Table 8.1). The minimum transfer time is 2 minutes, and hence, there is a buffer time of $10 - (6 + 2) = 2$ minutes.

8.2 CONSTRUCTION OF THE TIMED EVENT GRAPH

A timed event graph can be constructed from the line and synchronization data tables as introduced in Section 8.1. This construction consists of two parts: the derivation of a timed bipartite directed graph and the determination of the initial marking. A timed event graph is characterized by $\mathcal{G} = (\mathcal{P}, \mathcal{Q}, \mathcal{D}, \mathcal{M}_0, \mathcal{T})$; see Section 7.1. The first part of the construction will yield $\mathcal{P}, \mathcal{Q}, \mathcal{D}$, and $\mathcal{T}$, and the second part will yield $\mathcal{M}_0$.

Algorithm 8.2.1 (For timed bipartite graphs) *Input: the line and synchronization data files. Output: a timed bipartite graph.*

1. **Initialization.** *Number the rows of the line data file from 1 to n (actually, this has already been done in column 1). Set $i := 1$.*

2. **New line.** *For row i of the line data file, add a transition q_i and set $s := i$ (first line segment of new line).*

3. **Next line segment.** *If row $i + 1$ exists ($i + 1 \leq n$) and corresponds to the same line (rows i and $i + 1$ have equal entries in the second column), then add a transition q_{i+1}, a place $p_{i+1,i}$, an arc from q_i to $p_{i+1,i}$, and an arc from $p_{i+1,i}$ to q_{i+1}. Define the holding time of place $p_{i+1,i}$ as the sum of the run time and the dwell time of row i. Set $i := i + 1$ and go to step 3.*

4. **Line endpoint.** *Row i corresponds to the last line segment of a line. Add a place p_{si}, an arc from q_i to p_{si}, and an arc from p_{si} to q_s (s as defined in step 2). Define the holding time τ_{si} of place p_{si} as the sum of the run time and the dwell time of row i. If $i + 1 \leq n$, then set $i := i + 1$ and go to step 2; otherwise, go to step 5.*

5. **Number synchronization data.** *Number the rows of the synchronization data file from 1 to m (again, this has already been done in column 1). Set $k := 1$.*

6. **Synchronization.** *For the first (feeder) and second (connecting) line segment in row k of the synchronization data file, search the associated rows in the line data file. Suppose that these rows have numbers i and j, respectively. Add a place p_{ji}, an arc from q_i to p_{ji}, and an arc from p_{ji} to q_j. Define the holding time of place p_{ji} as the sum of the run time of row i in the line data file and the transfer time of row k in the synchronization data file. If $k < m$, then set $k := k + 1$ and go to step 6; otherwise, stop.*

The next step is to determine the initial marking. Let $d_i \in \{0, 1, 2, \ldots, T-1\}$ be the scheduled departure time associated with transition q_i as given in row i of the line data file. The quantity T is the cycle time (or period length) of the timetable. In this case $T = 60$ minutes. The following theorem gives an initial marking that is consistent with the timetable given in the line data file. The expression $\lceil a \rceil$ in the theorem refers to the ceiling function of the number a, and it equals the smallest integer number larger than or equal to a.

THEOREM 8.1 *Let $(\mathcal{P}, \mathcal{Q}, \mathcal{D}, \mathcal{T})$ be the timed bipartite graph given by the previous algorithm. Let $d \in \{0, \ldots T - 1\}^{|\mathcal{Q}|}$ be the scheduled departure times (modulo T) associated with the transitions q_i. An initial marking is denoted by the number*

m_{ji} *referring to the number of tokens in place* p_{ji} *with upstream transition* $q_i \in \mathcal{Q}$ *and downstream transition* $q_j \in \mathcal{Q}$, *and it is determined by*

$$m_{ji} = \left\lceil \frac{\tau_{ji} + d_i - d_j}{T} \right\rceil. \tag{8.1}$$

Proof. The initial marking is determined in such a way that the corresponding timed event graph can be executed with cycle time T and firing instants $d_i \pmod{T}$. Consider a pair of transitions q_i and q_j connected by an intermediate place p_{ji}. Assume that q_i fires at time instant $d_i \in \{0, \ldots, T-1\}$. At that moment a token is added to p_{ji}, which will be available at q_j at time instant $d_i + \tau_{ji}$. According to the timetable, q_j must fire at time instants $d_j + k \times T$, $k \in \mathbb{N}$, and so there must be a token available (i.e., being in the enabling state) in p_{ji} at these time instants. The token fired by q_i is only available after a number of cycles given by

$$m_{ji} = \min\{k \in \mathbb{N} | d_j + k \times T \geq d_i + \tau_{ji}\} = \min\left\{k \in \mathbb{N} \Big| k \geq \frac{d_i + \tau_{ji} - d_j}{T}\right\}.$$

The latter expression equals (8.1). Thus the initial marking of p_{ji} must contain at least this amount of tokens in order to enable the firing at the scheduled time instants. $\square$

Example 8.2.1 *Consider the railway system as introduced in Section 8.1. The number of transitions* $|\mathcal{Q}|$ *equals the number of rows in the line data file, so* $|\mathcal{Q}| = 24$. *The number of places* $|\mathcal{P}|$ *equals the sum of the number of rows in both the line and synchronization data files, i.e.,* $|\mathcal{P}| = 24 + 7 = 31$. *The timed event graph is shown in Figure 8.3. The four circuits associated with the four railway lines are clearly visible. The event graph is not strongly connected, as line 4, consisting of the transitions* $q_{21}, q_{22}, q_{23}, q_{24}$ *and the places in between these transitions, is not connected to the other lines. The reason is that Table 8.2 does not contain synchronization data with respect to line 4.*

The number of tokens (or trains) in each place can easily be determined. As an example, for place p_{21}, *with upstream transition* q_1 *and downstream transition* q_2, $m_{21} = \lceil (\tau_{21} + d_1 - d_2)/60 \rceil = \lceil (34 + 34 - 10)/60 \rceil = 1$. *Likewise, as two other examples,* $m_{32} = \lceil (38 + 10 - 49)/60 \rceil = 0$ *and* $m_{43} = \lceil (70 + 49 - 38)/60 \rceil = 2$.

8.3 STATE SPACE DESCRIPTION

In the chapter on Petri nets, the transition from a timed event graph to a max-plus recurrence relation was described. Generally, the state vector describes the state of the system, and in the particular case of timed event graphs, it describes the moments of firing of the transitions in the event graph. Before doing so here, we will first simplify the event graph somewhat, leading to a reduction of the dimension of the state vector. Suppose that a state description is given by

$$x(k+1) = \bigoplus_{l=0}^{M} A_l \otimes x(k-l), \tag{8.2}$$

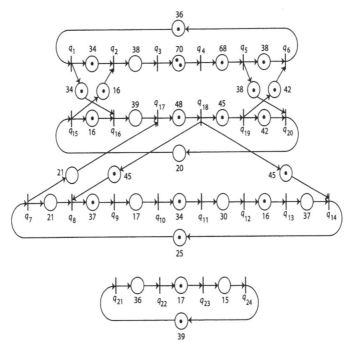

Figure 8.3: The timed event graph obtained by applying Algorithm 8.2.1 to the data from Tables 8.1 and 8.2. The numbers at the places indicate the sum of travel and waiting times from the upstream to the downstream transitions.

where $x(k) \in \mathbb{R}^n$, for all $k \geq -M$, and $A_l \in \mathbb{R}_{\max}^{n \times n}$, with $0 \leq l \leq M$. Then, by definition, the dimension of the state is n and the order of the system is M. The following algorithm will reduce the number of state variables (i.e., n), possibly at the expense of an increase of the order of the system.

Algorithm 8.3.1 (For state reduction) *Input: a timed event graph. Output: a reduced timed event graph.*

1. *If a transition q_j has only one upstream place p_{ji} and one downstream place $p_{kj} \neq p_{ji}$, then remove transition q_j and replace the places p_{ji}, p_{kj} by a place p'_{ki} with the following holding time and initial marking:*

$$\tau'_{ki} = \tau_{kj} + \tau_{ji}, \quad m'_{ki} = m_{kj} + m_{ji}.$$

Replace the original timed event graph by the new one. If still appropriate, repeat step 1; otherwise, go to step 2.

2. *For each transition q_j that has only one downstream place p_{kj}, which is not simultaneously an upstream place of q_j (loops are excluded from consideration), and that has multiple upstream places, do the following:*

- *For each upstream place p_{ji} of transition q_j, with output transition q_j and an input transition $q_i \neq q_j$, remove p_{ji} and add a place p'_{ki} with input transition q_i and output transition q_k with*

$$\tau'_{ki} = \tau_{kj} + \tau_{ji}, \ m'_{ki} = m_{kj} + m_{ji}.$$

- *Remove transition q_j and place p_{jk}. Replace the old event graph by the new one. If still appropriate, return to step 2; otherwise, stop.*

Let us give an example of this algorithm.

Example 8.3.1 *Consider the timed event graph in Figure 8.3. The reduced timed event graph at the moment that step 1 of Algorithm 8.3.1 has been executed completely is depicted in Figure 8.4. The reduced timed event graph when subsequently only transition q_8 has been removed, as described in step 2 of Algorithm 8.3.1, is depicted in Figure 8.5. The reduced timed event graph when the whole algorithm has been executed is depicted in Figure 8.6 (left). The original timed event graph of Figure 8.3 with twenty-four transitions and thirty-one places has been reduced to a timed event graph with seven transitions and fourteen places.*

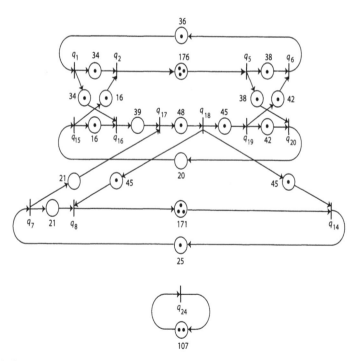

Figure 8.4: The reduced timed event graph after completely having carried out step 1 of Algorithm 8.3.1 starting from Figure 8.3.

Now that we have obtained an event graph (either the original or the reduced one), a state space description can be derived as already shown in Section 7.2. A

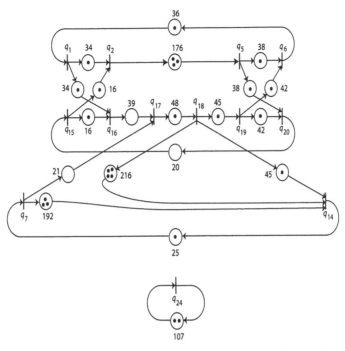

Figure 8.5: The reduced timed event graph when only transition q_8 has been removed according to step 2 of Algorithm 8.3.1 starting from Figure 8.4.

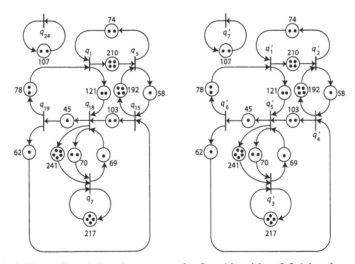

Figure 8.6: The reduced timed event graph after Algorithm 8.3.1 has been applied completely (left), and the same reduced timed event graph after the transitions have been relabeled (right).

	1	2	3	4	5	6	7
1		74_2				78_2	
2	210_4			192_4			
3			217_4		$70_2,241_5$		
4		58_1				62_1	
5	121_2		69_1	103_2			
6					45_1		
7							107_2

Table 8.3: The A_l matrices. The subscripts of the elements refer to the l concerned.

state variable corresponds to each transition of a timed event graph (for instance, the one in Figure 8.6 (left)). For transition q_j we define

$$x_j(k) = \bigoplus_{i \in \pi(j)} \tau_{ji} \otimes x_i(k - m_{ji}).$$

(Note that here $\pi(j)$ refers to the set of immediate upstream *transitions* with respect to transition q_j. In the previous chapter the notation $\pi(q_j)$ referred to the set of immediate upstream *places* of q_j. Generally, it will be clear from the context whether upstream places or transitions are meant.) Thus, the time behavior $x_j(k)$ at transition q_j is directly determined by the behavior of its upstream transitions. Recall that τ_{ji} refers to the holding time (sum of run time and transfer time) of the place between q_j and its upstream transition q_i and that m_{ji} is the number of tokens in this place. By introducing matrices A_l, $l \in \underline{M} \cup \{0\}$, where $M = \max_{i,j \in \underline{n}'} m_{ji}$, whose elements are defined by

$$[A_l]_{ji} = \begin{cases} \tau_{ji} & \text{if } m_{ji} = l, \\ \varepsilon & \text{otherwise}, \end{cases}$$

for all $i, j \in \underline{n}'$, where n' is the number of transitions in the reduced graph, we can now formally write for the state vector $x(k)$

$$x(k) = \bigoplus_{l=0}^{M} A_l \otimes x(k - l). \tag{8.3}$$

The matrices A_l corresponding to the graph of Figure 8.6 (right), which is identical to Figure 8.6 (left) apart from the numbering of the transitions ($q_1' := q_1$, $q_2' := q_5$, $q_3' := q_7$, $q_4' := q_{15}$, $q_5' := q_{18}$, $q_6' := q_{19}$, $q_7' := q_{24}$), are given in Table 8.3. In order to save space, these matrices for various l have been superimposed upon each other, but subscripts have been added to the holding times referring to the corresponding $l = 0, 1, \ldots, 5$. The empty spaces refer to elements ε. Note that in row 3, column 5 of Table 8.3 two elements appear; this is due to the fact that two places exist between transitions q_5' and q_3'. If the initial states $x(-1), x(-2), \ldots, x(-M)$ are given, then (8.3) determines the successive departure times. (Note that the graph of A_0 does not contain circuits, which is a sufficient condition for (8.3) to make sense as a recurrence relation.)

Railway systems usually operate according to a timetable demanding that a train is not allowed to depart before its scheduled departure time. A timetable can be incorporated in the max-plus linear system by addition of an inhomogeneous term,

$$x(k) = \bigoplus_{l=0}^{M} A_l \otimes x(k - l) \oplus d(k), \tag{8.4}$$

where $d(k) = (d_1(k), d_2(k), \ldots, d_{|n'|}(k))^\top$ is the vector of scheduled departure times. For a periodic timetable with period (or cycle time) T (i.e., a timetable in which every period of T time units the same processes are repeated) we have

$$d(k) = T^{\otimes k} \otimes d(0), \qquad k = 1, 2, 3, \ldots, \tag{8.5}$$

where $d(0)$ is the departure schedule for the initial trains. A necessary condition for a planned timetable to be feasible is that $d(k) \geq \bigoplus_{l=0}^{M} A_l \otimes x(k - l)$. This simply means that a train should have arrived and all connections secured before its scheduled departure time. The buffer time in the dwell/transfer time of a train connection between i and j is given by

$$r_{ji}(k) = d_j(k) - \bigoplus_{l=0}^{M} [A_l]_{ji} \otimes d_i(k - l). \tag{8.6}$$

If the departure times are given by (8.5), then the buffer time will be independent of k.

It is almost superfluous to state here that system (8.4) can be rewritten as a first-order system

$$\tilde{x}(k + 1) = \tilde{A} \otimes \tilde{x}(k) \oplus \tilde{d}(k + 1), \tag{8.7}$$

where $\tilde{x}$ and $\tilde{d}$ are augmented versions of the state vector x and departure time vector d in (8.4). This has been described in Section 4.5.

8.4 APPLICATION OF HOWARD'S ALGORITHM

The application of the policy iteration algorithm, as described in Chapter 6 and specifically in Section 6.3 for higher-order systems, to system (8.3) is left as an exercise for the reader. The result is the following. The system does not have a unique eigenvalue, which may come as no surprise since it is reducible. For the subsystem made up of transitions $q_1', q_2', \ldots, q_6'$ and the arcs connecting these transitions, to be called subsystem 1, the eigenvalue is $\lambda_1 = 54\frac{1}{4}$. For subsystem 2, consisting of q_7' with its only circuit, the eigenvalue is $\lambda_2 = 53\frac{1}{2}$. The eigenvectors for subsystems 1 and 2 are

$$\left(0, 13\frac{1}{4}, 25, 38\frac{1}{4}, 39\frac{3}{4}, 30\frac{1}{2}\right)^\top \quad \text{and} \quad 0, \tag{8.8}$$

respectively.

We now return to the original railway system as given in Figure 8.3. For that system the eigenvalue(s) and eigenvector(s) can be computed, just as above for the reduced railway system in Figure 8.6 (right). Another way of calculating these quantities is to make use of the eigenvalues and eigenvectors already obtained for the system in Figure 8.6 (right). The eigenvector is obtained by backward calculation starting from transitions for which the corresponding entry in the eigenvector is known. Let us work out this second method in some detail and consider transition q_{24} in Figure 8.3, which coincides with transition q'_7 in Figure 8.6 (right). The departure time of q_{24}, to be denoted by v_{24}, is set equal to the departure time of q'_7 as determined by the last entry of the second eigenvector in (8.8), i.e. $v_{24} = 0$. We can determine the departure time of q_{23}, to be denoted by v_{23}, by solving $(v_{23} + 15) \bmod 53\frac{1}{2} = v_{24}$. The number 15 in this equation refers to the travel and waiting time from q_{23} to q_{24}, and the number $53\frac{1}{2}$ represents the eigenvalue λ_2. Thus, we obtain $v_{23} = 38\frac{1}{2}$. Continuing in this way, with obvious notation, we get that v_{22} is determined by $(v_{22} + 17 - 53\frac{1}{2}) \bmod 53\frac{1}{2} = v_{23}$, resulting in $v_{22} = 21\frac{1}{2}$. The term $-53\frac{1}{2}$ in this expression (between the parentheses) is due to the token in the place between q_{22} and q_{23}. As a last example, let us determine v_4. It satisfies $(v_4 + 68 - 54\frac{1}{4}) \bmod 54\frac{1}{4} = v_5 = 13\frac{1}{4}$, the right-hand side being the second entry in the first eigenvector in (8.8), which results in $v_4 = 53\frac{3}{4}$. Recall that all departure times are measured in minutes past the full hour. The ultimate result is the generalized eigenvector of the max-plus model $\tilde{x}(k+1) = \tilde{A} \otimes \tilde{x}(k)$, with $\tilde{A}$ as in (8.7) corresponding to Figure 8.3 and given by

$$v = \left(0, 0, 38, 53\frac{3}{4}, 13\frac{1}{4}, 18\frac{1}{4}, 25, 46, 28\frac{3}{4}, 45\frac{3}{4}, 25\frac{1}{2}, 1\frac{1}{4}, 17\frac{1}{4}, 0, 38\frac{1}{4}, 7, 46, \right.$$

$$\left. 39\frac{3}{4}, 30\frac{1}{2}, 18\frac{1}{4}, 39, 21\frac{1}{2}, 38\frac{1}{2}, 0 \right)^{\top}.$$

The corresponding cycle-time vector is a vector with twenty-four entries, the first twenty having the value $54\frac{1}{4}$ and the last four having the value $53\frac{1}{2}$. It then follows that for all $k \geq 0$

$$\tilde{A} \otimes (k \times \eta + v) = (k+1) \times \eta + v.$$

8.5 EXERCISES

1. Suppose that a railway system is characterized by a timed event graph in which each place has at least one token. The corresponding timetable is given by an equation like (8.4). Each place with more than one token is now split up into a series of auxiliary places with one token each (and auxiliary transitions in between). Show that if one derives a max-plus model from this latter event graph, then one obtains a first-order model like (8.7).

2. Apply Howard's algorithm to the higher-order system given in (8.3) and show that (8.8) is correct.

3. Show that $\tilde{d}(k+1) \geq \tilde{A} \otimes \tilde{d}(k)$ (see (8.7) for the definition of these quantities) is a necessary condition for the feasibility of the timetable.

4. Check the eigenvalue(s) of the twenty-four-dimensional system as given in Section 8.4.

8.6 NOTES

An interactive software tool called PETER [46], which is an acronym for Performance Evaluation of Timed Events in Railways, has been developed by Ortec Consultants BV, The Netherlands, in close collaboration with Delft University of Technology. PETER is a recently developed analysis tool, mainly based upon the max-plus algebra approach, that quickly assesses network performance indicators of large models (thousands of variables) in a deterministic setting corresponding to design times used in timetable construction (see [46] and [77]).

If one wants to combine the models of the railway nets of two neighboring countries, such as Germany and the Netherlands, into one larger model, a helpful tool for its construction is that of z-transforms and transfer matrices (see [5]) in a max-plus setting.

Results of the full Dutch network have been described in [16] and [82]. The example central to this chapter appeared in [47]. Though the max-plus approach to the design of timetables provides specific insights, the practical design of the Dutch railway timetable has been based, up till now, primarily on mixed-integer techniques; see [80]. Section 8.4 relies on [81].

This book concentrates on applications with respect to timetables. Applications in the area of network calculus can be found in [59]. Applications in the production environment, like assignment problems, are reported in [67], for example.

The following story comes from one of the authors: "When I invited myself to the board of the Dutch Railway Company in order to explain about the advantages of a max-plus approach to timetable design, I was received very kindly and asked to tell my story, which I did in simple words, with hardly a formula, if any. The atmosphere corresponded with my experience with introductions of other mathematical disciplines for different groups of people: mathematicians are generally considered to be gentle persons and a nondangerous, nonaggressive, and sometimes amusing species. Not surprisingly, nothing happened afterwards. No further active interest was shown. This changed when I got colleagues from the department of civil engineering of our university interested. Since that time, they function as a catalyst, and now a fruitful collaboration with the railway company has been established."

Chapter Nine

Delays, Stability Measures, and Results for the Whole Network

This chapter is a follow-up to the previous one. Once a timetable is given, we are interested in its sensitivity with respect to disturbances in the system. A question that came up during one of the discussions at the Dutch railway headquarters was, how many minutes can all changeover times be increased such that a timetable with a period of sixty minutes still can be maintained? The underlying reason for this question was that an increase in age of the average passenger is expected during the coming years due to the baby boom after World War II. Older passengers walk more slowly, hence the idea to consider increasing the changeover times.

In the first part of this chapter, Section 9.1, the operability of the timetable is discussed when disturbances are considered. Various notions of stability or robustness are discussed as well.

In the previous chapter, the modeling and some results were explained by means of a submodel of the Dutch railways. Some results for the whole network, including some discussion with respect to stability, will be given in Section 9.2. Finally, in Section 9.3, some related modeling issues are discussed, such as the integration of lines with different frequencies into the same model.

9.1 PROPAGATION OF DELAYS

An important consideration in timetable analysis is how initial delays propagate through the network when all connections remain secured (i.e., the model does not change). This theory is most easily explained when the model is given as a first-order recurrence relation similar to (8.7). Hence, in this section we will confine ourselves to such models.

In the following, we assume that only one delay will occur, and that, without loss of generality, it will take place with respect to the initial value $x(0)$. The aim is then to determine which trains will be affected by this initial delay.

The starting point is the model as given in (8.7), where, for simplicity of notation, we leave out the tilde symbols $\tilde{}$. From this model we derive the following:

$$
\begin{aligned}
x(k+1) &= A \otimes x(k) \oplus d(k+1) \qquad\qquad (9.1)\\
&= A \otimes (A \otimes x(k-1) \oplus d(k)) \oplus d(k+1)\\
&= A^{\otimes 2} \otimes x(k-1) \oplus A \otimes d(k) \oplus d(k+1)\\
&= A^{\otimes 2} \otimes x(k-1) \oplus d(k+1).
\end{aligned}
$$

The last relation holds because $d(k+1) \geq A \otimes d(k)$ for a feasible timetable. With an induction argument we conclude that

$$x(k+1) = A^{\otimes(k+1)} \otimes x(0) \oplus d(k+1). \qquad (9.2)$$

If there are no initial delays, then $x(0) = d(0)$ and (9.2) can be written as

$$x(k+1) = A^{\otimes(k+1)} \otimes d(0) \oplus d(k+1) = d(k+1).$$

Relation (9.2) can be used to determine the propagation of delays in the following way. Let a first train be delayed, so that for a certain j, $x_j(0) > d_j(0)$. Furthermore, assume this is the only train that is initially delayed; in other words assume $x_i(0) = d_i(0)$ for $i \neq j$. From (9.2) it follows that the initial delay on node j causes a delay for the $(k+1)$st train departing from node i if

$$\bigoplus_{l=1}^{n} \left[A^{\otimes(k+1)}\right]_{il} \otimes x_l(0) > d_i(k+1),$$

where n is, as usual, the dimension of the state. Because of our assumption that $x_i(0) = d_i(0)$ for $i \neq j$, it follows from (9.2) that the $(k+1)$st departure at node i is delayed because of a delay in the initial departure at node j if

$$\left[A^{\otimes(k+1)}\right]_{ij} \otimes x_j(0) > d_i(k+1). \qquad (9.3)$$

In this way, we obtain the set of all delayed trains. Observe that the matrices $A^{\otimes k}$, $k = 1, 2, \ldots$, can be calculated in advance and that to determine the propagation of an initial delay at node j we only need the jth column of these matrices.

From (9.3) it follows that after k_j^* steps, where k_j^* is given by

$$k_j^* = \min\left\{k \in \mathbb{N} : \left[A^{\otimes(k+1)}\right]_{ij} \otimes x_j(0) \leq d_i(k+1)\ \forall i \in \underline{n}\right\},$$

the initial delay is out of the system. This k_j^* resembles but is different from the transient time as introduced in Chapter 3, since k^* depends also on the timetable. Note that in the case of a tight timetable, $k^* = +\infty$.

9.1.1 Stability and traffic rate

The eigenvalue, or equivalently the maximum cycle mean, determines the highest frequency for which a timetable can be designed and subsequently operate. The eigenvalue equals the average interdeparture time of a train on a critical circuit. A critical circuit can hence be viewed as the slowest circuit in the network. The mean cycle time of such a circuit determines the minimum possible cycle time, or equivalently the highest frequency, for the entire system. In more romantic terms, the critical circuit determines the heartbeat of the system.

In practice it is not advisable to operate according to a tight timetable with minimum cycle time. A timetable should be able to compensate for slight disturbances or disruptions in run and waiting times. With a tight timetable the consequences of a delay (i.e., its propagation) on a critical circuit will never die out. For that reason, margins and buffer times will be incorporated in the timetable, which will increase the minimum cycle time. Another very practical reason not to choose a tight timetable is that the period T should be such that the departure times can be easily remembered by heart (e.g., $T = 60$ or $T = 30$ minutes). The sensitivity with respect to delays in a transportation system leads to the subject of stability and robustness.

DEFINITION 9.1 *A scheduled train service system is called* stable *if any initial delay (assumed to be finite) settles in finite time.*

THEOREM 9.2 *Consider system (9.1), with the inhomogeneous terms determined by (8.5). This system is stable if and only if $\lambda < T$, where λ is the eigenvalue defined by $A \otimes v = \lambda \otimes v$ and T is the cycle time of the timetable as defined by means of (8.5).*

The proof of this theorem is a direct consequence of the theory treated in Section 3.1 and is left as an exercise. The traffic rate, denoted by ρ, is defined by λ / T. It is a performance indicator denoting the trade-off between the maximum performance ($T = \lambda$) under ideal circumstances and robustness. Obviously, a stable system requires $0 \le \rho < 1$, whereas $\rho = 1$ corresponds to the saturated case.

Remark. The above theorem can easily be rephrased for systems described by (8.4). In that case λ is defined implicitly by $\mathcal{A}(\lambda^{\otimes -1}) \otimes v = v$, where

$$\mathcal{A}(\lambda^{\otimes -1}) \overset{\text{def}}{=} \bigoplus_{l=0}^{M} A_l \otimes \lambda^{\otimes -l} \, ;$$

for the definition of a negative exponent, see Section 1.1.

9.1.2 Stability margin

The stability margin Δ is a measure of robustness of the system, that is, how stable the system is.

DEFINITION 9.3 *The stability margin Δ with respect to a set of parameters is the maximum amount of time that can be added to all these parameters simultaneously such that the corresponding network is still operable with the given period T; that is, the eigenvalue of the new system matrix with the maximum Δ included equals T.*

Please be aware of the fact that with Δ added to the system parameters, the departure times are determined by the eigenvector of the new system matrix and will generally be different from the original ones.

If one talks about the stability margin without any further specification, it is assumed that *all* parameters of the system (travel times, waiting times, ...) are increased by the same amount. One must be careful with the interpretation of "all parameters of the system". If the system is identified with the mathematical model, then the full state model can be meant as well as the reduced state model (in which parameters, or travel times, are linked together into new single parameters). Further, note that a first-order model, if derived from a higher-order model, contains many structural unit elements (i.e., e's), and one probably does not want to increase these structural unit elements by Δ. No closed-form expression for Δ is known to exist if the system is given by (8.4). Numerically, one can determine Δ by making use of the fact that the eigenvalue is nondecreasing with respect to Δ. For first-order systems, however, a closed-form expression does exist, as shown by the following theorem, whose proof is left as an exercise.

THEOREM 9.4 *The stability margin Δ for the first-order system (9.1), with respect to all parameters, is determined by $\Delta = T - \lambda$, where λ is the solution of $A \otimes v = \lambda \otimes v$, with v finite.*

Compare this theorem with exercise 3 in Section 9.4.

9.1.3 Recovery times

DEFINITION 9.5 *Consider the max-plus linear system (9.1). The entry r_{ji}^+ of recovery matrix R^+ is defined as the maximum delay of $x_i(0)$ such that $x_j(k)$ is not delayed for any $k \geq 1$.*

Remark. Please note the difference between r_{ji} as introduced in (8.6) and r_{ji}^+ as introduced in Definition 9.5. In the definition of r_{ji} only direct connections are considered, whereas in the definition of r_{ji}^+ composite connections are considered as well.

THEOREM 9.6 *For system (9.1) the elements of the recovery matrix R^+ are given by*

$$r_{ji}^+ = d_j - d_i - [A \otimes T^{\otimes-1}]_{ji}^+, \tag{9.4}$$

where the vector d equals the vector $d(0)$ as given in (8.5) and where the notation $[A \otimes T^{\otimes-1}]_{ji}^+$ refers to the ji-th element of the matrix $[A \otimes T^{\otimes-1}]^+$. If in the graph of A no path exists from node i to node j, then $r_{ji}^+ = +\infty$.

Proof. Suppose a delay of r occurs with respect to departure time d_i. In order for the delay not to have any influence on the next departure times,

$$A^{\otimes k} \otimes \begin{pmatrix} d_1 \\ \vdots \\ d_{i-1} \\ d_i + r \\ d_{i+1} \\ \vdots \\ d_n \end{pmatrix} \leq \begin{pmatrix} d_1 \\ \vdots \\ d_{i-1} \\ d_i \\ d_{i+1} \\ \vdots \\ d_n \end{pmatrix} \otimes T^{\otimes k}, \qquad k = 1, 2, \ldots,$$

must hold. The jth component reads

$$\max\left([A^{\otimes k}]_{j1} + d_1, \ldots, [A^{\otimes k}]_{ji} + d_i + r, \ldots, [A^{\otimes k}]_{jn} + d_n\right) \leq d_j + k \times T,$$

and, hence in particular $[A^{\otimes k}]_{ji} + d_i + r \leq d_j + k \times T$. Therefore, the maximum delay that does not cause a follow-up delay at node j for any k equals $\min_k(d_j - $

$d_i - [A^{\otimes k}]_{ji} + k \times T)$, equivalently, $d_j - d_i - [A \otimes T^{\otimes -1}]_{ji}^{+}$, which by Definition 9.5 can also be denoted by r_{ji}^{+}. This proves the theorem. □

Please note that this proof has a close relationship with the principal solution as introduced in Section 2.3. This theorem can also be formulated and (directly) proved with respect to the system (8.4). The term $-[A \otimes T^{\otimes -1}]_{ji}^{+}$ in (9.4) is then replaced by $-(\mathcal{A}(T^{\otimes -1}))_{ji}^{+}$. See the exercises at the end of the chapter.

Example 9.1.1 *The recovery matrix R^{+} of size 24×24, with $T = 60$, of the example railway network constructed from Tables 8.1 and 8.2, presented as a Petri net in Figure 8.3, is given in Table 9.1. Due to space limitations, this table presents the recovery matrix in two parts. The top part lists the first twelve columns of R^{+} and the bottom part the last twelve columns. The dots in this table represent the number $+\infty$. As an illustration, consider the subnetwork of line 4, consisting of the line segments 01, 02, 51, and 52 in Table 8.1; this corresponds to the 4×4 matrix at the right bottom of Table 9.1.*

	1	2	3	4	5	6	7	8	9	10	11	12
1	56	74	73	34	32	30	59	82	82	82	68	68
2	2	50	49	10	8	32	35	58	58	58	44	44
3	3	1	50	11	9	33	36	59	59	59	45	45
4	42	40	39	50	48	72	75	98	98	98	84	84
5	44	42	41	2	50	74	77	100	100	100	86	86
6	26	44	43	4	2	56	29	52	52	52	38	38
7	38	86	85	46	44	68	23	23	23	23	9	9
8	24	72	71	32	30	54	0	23	23	23	9	9
9	24	72	71	32	30	54	0	0	23	23	9	9
10	24	72	71	32	30	54	0	0	0	23	9	9
11	38	86	85	46	44	68	14	14	14	14	23	23
12	38	86	85	46	44	68	14	14	14	14	0	23
13	38	86	85	46	44	68	14	14	14	14	0	0
14	29	77	76	37	35	59	14	14	14	14	0	0
15	30	48	47	8	6	60	33	56	56	56	42	42
16	2	50	49	10	8	32	35	58	58	58	44	44
17	2	50	49	10	8	32	5	28	28	28	14	14
18	23	71	70	31	29	53	26	49	49	49	35	35
19	24	72	71	32	30	54	27	50	50	50	36	36
20	25	43	42	3	1	55	28	51	51	51	37	37
21	.	.	.	.	.	.	.	.	.	.	.	.
22	.	.	.	.	.	.	.	.	.	.	.	.
23	.	.	.	.	.	.	.	.	.	.	.	.
24	.	.	.	.	.	.	.	.	.	.	.	.

	13	14	15	16	17	18	19	20	21	22	23	24
1	68	68	56	54	54	33	32	61	.	.	.	.
2	44	44	2	30	30	9	8	7	.	.	.	.
3	45	45	3	31	31	10	9	8	.	.	.	.
4	84	84	42	70	70	49	48	47	.	.	.	.
5	86	86	44	72	72	51	50	49	.	.	.	.
6	38	38	26	24	24	3	2	31	.	.	.	.
7	9	9	38	36	36	15	44	43	.	.	.	.
8	9	9	24	22	22	1	30	29	.	.	.	.
9	9	9	24	22	22	1	30	29	.	.	.	.
10	9	9	24	22	22	1	30	29	.	.	.	.
11	23	23	38	36	36	15	44	43	.	.	.	.
12	23	23	38	36	36	15	44	43	.	.	.	.
13	23	23	38	36	36	15	44	43	.	.	.	.
14	0	23	29	27	27	6	35	34	.	.	.	.
15	42	42	30	28	28	7	6	5	.	.	.	.
16	44	44	2	30	30	9	8	7	.	.	.	.
17	14	14	2	0	30	9	8	7	.	.	.	.
18	35	35	23	21	21	30	29	28	.	.	.	.
19	36	36	24	22	22	1	30	29	.	.	.	.
20	37	37	25	23	23	2	1	30	.	.	.	.
21	.	.	.	.	.	.	.	.	13	10	5	2
22	.	.	.	.	.	.	.	.	3	13	8	5
23	.	.	.	.	.	.	.	.	8	5	13	10
24	.	.	.	.	.	.	.	.	11	8	3	13

Table 9.1: The 24×24 recovery matrix, presented in two blocks of the first and second twelve columns, respectively.

The elements in the recovery matrix include the buffer time between two successive line segments. As an example, $r_{22,21}^+ = d_{22} - d_{21} - (A(T^{\otimes -1}))_{22,21}^+ = 59 - 20 - 36 = 3$. This implies that the train on line 4, segment 01 may be delayed for three minutes without affecting the departure time for line 4, segment 02. Note that $r_{21,21}^+ = r_{22,22}^+ = r_{23,23}^+ = r_{24,24}^+ = 13$. This can be explained as follows. The total number of tokens of line 4 in Figure 8.3 equals two, the sum of all holding times of line 4 is 107. Hence, $r_{ii}^+ = d_i - d_i - (107 - 2 \times 60) = 13$ for $i = 21, 22, 23$, and 24.

The above analysis assumes that only one train is initially delayed. A more general setup is the following. The vector of maximal (i.e., the latest) departure times such that the trains can just meet the timetable at their subsequent departures is given by the principal solution $x^*(A, d)$; see Theorem 2.11. Indeed, $x^*(A, d)$ solves the maximization problem $\max_x A \otimes x \leq d$. In the same vein, the vector of latest departure times such that the timetable can be met k transitions later is obtained from

$$z(k) \stackrel{\text{def}}{=} x^*(A^{\otimes k}, d(k)),$$

for $k \geq 1$. Thus, $z(k) - d(0)$ is the vector denoting the maximal initial delay that can be compensated after k transitions or, equivalently, the maximal initial delay that does not propagate beyond k transitions.

9.2 RESULTS FOR THE WHOLE DUTCH INTERCITY NETWORK

In this section some results for the whole Dutch intercity network, as of the year 2001, will be given. These results, as well as the figures to be presented, have been produced by means of the software package PETER (introduced in the notes section of Chapter 8). The network consists of 19 lines, serving 70 stations. The max-plus model consists of 317 departure events (i.e., nodes) and 361 line segments (i.e., arcs), including 44 connections (i.e., synchronization constraints) between trains. A minimum of 112 trains is necessary to cover all intercity train circulations. The model contains 137 tokens, which is a measure for the problem dimension.

Remark. In addition to intercity trains, there are two other types of trains, namely, express and local trains. Intercity trains only stop at main stations and are meant for the through traffic. Local trains stop at all stations and, as the name suggests, are meant for local distances. Express trains are somewhere between intercity and local trains. The model with all train types has a minimum of 441 trains. If, within this model, two types of trains (e.g., intercity and express trains) use the same track, then this track will be represented by two arcs in the corresponding event graph.

The event graph of the intercity network happens to consist of a number of maximal strongly connected subgraphs. Each of these subgraphs has a critical circuit. The critical circuit with the largest circuit mean is given in Figure 9.1 (the black links) together with the corresponding maximal strongly connected subgraph (white links). The remaining links (in gray) refer to the other tracks of the intercity network. This critical circuit, from Groningen to Schiphol (international

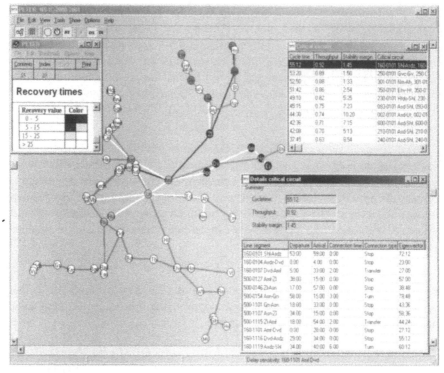

Figure 9.1: The intercity network, the most critical circuit (black lines) with the associated subgraph (white lines). The delay sensitivity is also indicated: the darker the station, the smaller the recovery time.

airport) and back, consists of parts of two lines, from Groningen to Amersfoort (changeover) and from Amersfoort to Schiphol. The cycle time is 55:12. This notation refers to 55 minutes and 12 seconds. The train network is operated according to an hourly timetable, giving a traffic rate of 92 percent. The stability margin Δ equals 1:45 minutes. In the table labeled "critical circuits" included within Figure 9.1, similar data is given for other maximal strongly connected subgraphs (which have a circuit mean less than 55:12).

Remark. At first sight the above stability margin may come as a surprise. One might expect that Δ should be equal to, or even be larger than, 60:00 - 55:12 = 4:48 minutes. The explanation why this is not true is that in the nonreduced event graph, where tracks are represented by places, roughly two out of three places do not contain tokens. The travel times of all tracks, also those corresponding to places without tokens, is increased by the stability margin. The cycle time is defined as the total travel time divided by the number of tokens of the circuit concerned.

The recovery matrix R^+ has also been computed. All trains on the critical circuit have a recovery time of twenty-four minutes. This means that if a train in a certain direction is delayed by twenty-four minutes, the next train in the same direction can

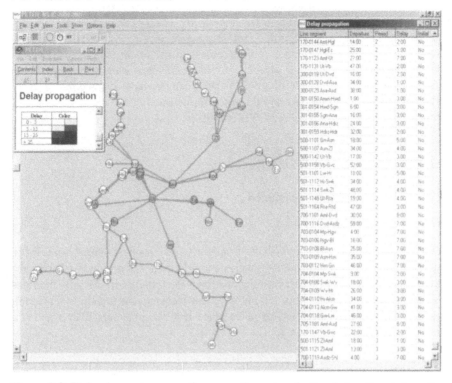

Figure 9.2: Delay impact propagation over the network: the darker the station, the longer the delay.

still leave on time without upsetting other connections. Figure 9.1 also visualizes the delay sensitivity of the train leaving Amersfoort in the direction of Amsterdam. Delay sensitivity, related to rows of R^+, refers to the amount of time incoming trains and other preceding trains can be late such that the train mentioned (departure Amersfoort, direction Amsterdam) can still leave on time with all the synchronizations secured. In the figure, the darker the stations are, the smaller the recovery time. The preceding stops of this train are most critical. A delay at Apeldoorn has zero recovery time (in Amersfoort). This robustness is gradually increased over the preceding stations to a four minute recovery time in Enschede. The preceding stops of feeder trains are the next critical.

Remark. Above, we discussed recovery times, related to the diagonal elements of R^+ and delay sensitivity, related to the rows of R^+. A third notion is the *delay impact* of a train, which is defined as the consequences for subsequent train sequences. The latter notion is related to the columns of R^+.

More detailed insight in the interconnectedness of the intercity train services is obtained by studying the delay propagation of initial delays. As an example, we consider a scenario in which, during one hour, each of the ten departing intercity trains from Utrecht has a ten minute departure delay. Figure 9.2 shows the resulting delay propagation. In total, forty-three stations are affected, including the terminal stations of Groningen, Leeuwarden, and Nijmegen, where the delays continue to propagate in the reverse direction (unless reserve trains are dispatched). The delays also reach five other terminal stations. However, at these stations enough buffer time is available to guarantee that the further train circulation is according to the timetable. Apart from the ten initial trains, thirteen additional trains get delayed, with an average departure delay of 5:14 minutes.

9.3 OTHER MODELING ISSUES

9.3.1 Features that can be included

Merging and splitting of trains

Merging and splitting of trains occurs frequently. Two trains, coming from different directions, are merged and continue as a single train. Or, a train is split up into two parts, each leaving for different destinations. Such events are trivially modeled by means of event graphs. (For the merging, for instance, the transition representing the station where the merging takes place has two upstream places and one downstream place.)

Different frequencies

The term *different frequencies* means that on certain tracks the frequency of the trains is higher than on others. In the Netherlands, for instance, most of the intercity trains have a frequency of one per half hour, whereas many local trains only go once per hour. This is easily modeled in max-plus algebra and will be illustrated by means of an example.

Example 9.3.1 *A layout of railway tracks is schematically shown in Figure 9.3 as a Petri net. The system consists of four circuits and three stations $S_i, i \in \underline{3}$, depicted by transitions $q_i, i \in \underline{3}$. There is one circuit between each of the three stations and one circuit connecting station S_1 with suburb stations (the latter are not explicitly indicated). One train runs on each of these circuits. These trains are symbolized by tokens, and at a particular moment they are situated as given in the figure. The travel times (with the dwell times included) are given along the tracks.*

It is required that the frequency along the track from S_1 serving the suburbs is twice as high as the frequency of the trains along the three other circuits. The arrivals of trains at S_1 from the high-frequency track should alternately give connections (i.e., possibilities for changeovers) for trains arriving from and leaving in the directions of S_2 and S_3, respectively. At stations S_2 and S_3 changeover connections should also be secured. It is claimed now that the following equations are a max-plus model for this layout and constraints:

$$
\begin{aligned}
x_1(k+1) &= \max(x_2(k) + 26, x_3(k+1) + 19), \\
x_2(k+1) &= \max(x_1(k) + 26, x_4(k+1) + 24), \\
x_3(k+1) &= \max(x_1(k) + 25, x_4(k) + 20), \\
x_4(k+1) &= \max(x_2(k) + 22, x_3(k+1) + 27).
\end{aligned}
\tag{9.5}
$$

The quantity x_1 refers to the departure time of the trains in the directions of the suburbs and S_2; x_2 refers to the departure time in the directions of the suburbs and S_3; and x_3 and x_4 refer to the departure times at stations S_2 and S_3, respectively. The essential feature of this modeling is that the high-frequency circuit appears twice in the model (by means of x_1 and x_2).

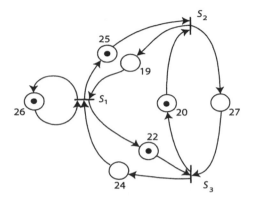

Figure 9.3: Petri net for Example 9.3.1.

In more general situations, with more different frequencies $f_i, i \in \underline{s}$, one takes the least common multiple of $f_i^{-1}, i \in \underline{s}$, where the unit in which f_i^{-1} is supposed to be expressed in minutes, with f_i^{-1} being the number of minutes it takes between two subsequent departures. The submodel related to one particular frequency f_i appears in the total model as many times as f_i^{-1} fits in the least common multiple just defined.

Personnel shifts

Personnel in the train (engine drivers, guards, and possibly waiters) do not necessarily have their duties on the same train throughout the day. Quite often they change trains. If such shifts are periodic, then they can in principle be included in max-plus modeling. For this modeling, it does not matter whether the tokens or places represent trains or persons. It is almost unnecessary to state that the model will become high dimensional.

Maximum waiting times

The typical equation for the departure time is

$$x_i(k+1) = \max_{j \in \underline{n}}(x_j(k) + a_{ij}), \qquad i \in \underline{n}.$$

Suppose that one wants to introduce a maximum dwell time at the station (such that the passengers already in the train do not have to wait too long). Then one can write

$$x_i(k+1) = \min\left(\max_j(x_j(k) + a_{ij}), x_l(k) + a_{il} + b_l \right), \qquad (9.6)$$

where S_l is the station from which the departing train comes originally. Departure will take place within b_l minutes of the arrival time, irrespective of whether all other trains have arrived for the transfer of passengers. The right-hand side of (9.6) is a so-called min-max-plus expression to be dealt with in Section 12.1. In reality (9.6) is only useful if one can reasonably assume that the train from S_l arrives fewer than b_l minutes late. If one has reason to believe that this train will have a delay of more than b_l minutes, then there is no reason to wait at all.

9.3.2 Limitations in the modeling

Optimal allocation of trains

In Chapter 0 a simple railway network was studied, where in Section 0.4 trains were added to the system in order to decrease the average interdeparture time, equivalently, to increase the frequency of the departures of the trains. We can also pose the question of how to position the existing trains on the network so as to achieve the highest possible frequency. One might consider the idea of taking away one train on a noncritical circuit and adding it to the critical circuit. The number of trains on the critical circuit is increased by one, thus this circuit may no longer be critical. In the latter case, another circuit, originally noncritical, has become critical. One might ask for the optimal allocation of a given number of trains in a network in order to achieve the highest frequency (or, in terms of the corresponding first-order max-plus algebra model, in order to achieve the smallest possible eigenvalue of the system matrix). This problem of the optimal allocation of trains is not easily solved, unfortunately (integer programming techniques seem to be more appropriate). One can in principle try all possible allocations (a huge, but finite number of possibilities) and calculate the eigenvalue for each of its corresponding models. Let us reconsider Example 9.3.1 for that purpose.

Example 9.3.2 *Consider the system as given in the Petri net of Figure 9.3 with corresponding model (9.5). By now it is a matter of routine to calculate the average interdeparture time. (If (9.5) is symbolically written as $x(k + 1) = A_0 \otimes x(k + 1) \oplus A_1 \otimes x(k)$, then this interdeparture time equals the eigenvalue of $A_0^* \otimes A_1$.) It turns out that $\lambda = 51$. One has some freedom to reallocate trains. Each circuit in Figure 9.3 has one token and one or two places. In each circuit with two places we can move the only token of this circuit to the other, empty place. Altogether this leads to eight different possibilities for the allocation of the four trains. By working through each of these possibilities, one finds that if the token in the circuit between S_1 and S_3 is moved to the other place (and not moving the remaining tokens), then $\lambda = 47$, and this is the minimum λ possible.*

A lower bound for the eigenvalue for the optimal allocation of trains can be given. See Exercise 4 of Chapter 13.

Ordering issues

Suppose that two trains must cross at a station with only one platform. The question is, if the trains arrive almost simultaneously, which one should go first and which one should wait? Especially during the design phase of a timetable, the order of these trains should be considered. Apart from the design phase, such questions occur also in practice when there are stochastic disturbances in the arrival times. One is undoubtedly tempted to prefer the first arriving train to stop at the station first, so that the last arriving train must wait. Suppose the arrival times of the arriving trains (which came from stations S_i and S_j, respectively) at station S_l are given by

$$x_i(h) + a_{li} \quad \text{and} \quad x_j(k) + a_{lj}$$

if a possible interaction is not considered. The arrival time of the first train is given by

$$\min(x_i(h) + a_{li}, x_j(k) + a_{lj}), \tag{9.7}$$

and the arrival time of the last one is given by

$$\max(x_i(h) + a_{li}, x_j(k) + a_{lj}).$$

One could continue the modeling by posing

$$x_r(k+1) = \max\{\ldots, \min(x_i(h) + a_{li}, x_j(k) + a_{lj}) + a_{rl}, \ldots\}.$$

This means that the first arriving train at station S_l, irrespective of whether it came from S_i or S_j, will aim for station S_r. This is in conflict with the fact that trains usually have fixed routes. This also would mean that passengers would have to change trains at S_l depending on whether their train arrives first or last. There is no easy way to resolve the conflicting issues of fixed routes of trains and the flexibility of the first arriving train having priority over the second, not knowing a priori which train will be the first to arrive. With the introduction of an equation similar to expression (9.7), one leaves the realm of max-plus algebra and enters that of min-max-plus algebra. The latter is the subject of Chapter 12.

9.4 EXERCISES

1. In Section 9.1.1 it was claimed that if the period of a timetable is equal to the minimum possible cycle time, a delay on a critical circuit will never die out. Prove this. As a specific example assume that a railway system consists of two stations, S_1 and S_2, with $a_{21} = 1$, $a_{12} = 3$, and $a_{11} = a_{22} = \varepsilon$. Assume furthermore that $d(k) = (1 + 2k \quad 2k)^\top$. If initially one has $x(0) = (1\frac{1}{2} \quad 0)^\top$, as opposed to $x(0) = d(0) = (1 \quad 0)^\top$, will the delay of half a minute ever die out?

2. Prove Theorem 9.2.

3. Consider (8.7). Assume that the eigenvalue λ of $\tilde{A}$ is strictly smaller than the period T. Assume furthermore that the communication graph of $\tilde{A}$ has a unique critical circuit. Prove that the stability margin Δ with respect to all elements of $\tilde{A}$ except for the structural e's satisfies

$$T - \lambda \le \Delta \le (T - \lambda)\frac{m}{m - \overline{m}},$$

where $\overline{m}$ is the number of nodes in the critical circuit related to the structural e's and m is the total number of nodes in the critical circuit.

4. Prove for any pair (i, j) that, $r_{ji} \geq r_{ji}^+$.

5. Prove Theorem 9.4. Explain why the generalization of this theorem does not hold for systems with higher-order terms, i.e., of the kind (8.4). This generalization would read $\Delta = T - \mu$, where $\mathcal{A}(T^{\otimes - 1}) \otimes v = \mu \otimes v$.

6. Consider the Petri net in Figure 9.3 and the corresponding model (9.5). Show that the maximum cycle mean equals 51. As shown in Section 9.3.2, it is possible by means of a different token distribution (the number of tokens, or trains, does not change) to get a smaller maximum cycle mean. What would be the worst possible token distribution in terms of getting the largest possible maximum cycle mean? (Answer: 102.)

7. Show that for system (8.4) the elements of the recovery matrix R^+ are given by

$$r_{ji}^+ = d_j - d_i - (\mathcal{A}(T^{\otimes - 1}))_{ji}^+,$$

where the vector d is given by (8.5). If q_i and q_j are not connected by a path, then $r_{ji}^+ = +\infty$.

9.5 NOTES

Results of the full Dutch railway network have been described in [16] and [82]. The notion of *recovery matrix* was introduced in [35]. Figures 9.1 and 9.2 have been produced with PETER. This package helped us to routinely construct (or, rather, fill) large A matrices with travel times. In the beginning stages, our PhD students Braker and Subiono did not have PETER at their disposal, and they extracted the travel times, one by one, from the annual timetable book as sold by the railway company. Braker thus filled in a 79×79 matrix with numerical data and Subiono a 341×341 matrix. Fortunately, there were many ε's, but many years later, they still feel unhappy when they think of all this work . . . .

Chapter Ten

Capacity Assessment

This chapter illustrates the application of max-plus algebra to models with sharing of and competition for resources. Section 10.1 will describe the occupation of a railway track (being the resource) by two types of trains. Slow and fast trains alternately use the track. The heaps of pieces approach, as introduced in Section 1.3, provides useful insights. Section 10.2 will deal with a real-life study in which the competition for resources stems from the situation in which a double-track railway line passes through three tunnels, each of them essentially functioning as a single-track section. Section 10.2.1 will deal with a stylized example, and Section 10.2.2 will provide a max-plus analysis of the capacity of the line containing three tunnels in total.

10.1 CAPACITY ASSESSMENT WITH DIFFERENT TYPES OF TRAINS

Consider a single track from station S_0 via S_1 to S_2. As a specific example, S_0 equals Rotterdam CS, S_1 equals Delft, and S_2 equals Den Haag HS; see Figure 8.1 for the geographical location of these stations. The track from S_0 to S_1 is called track 1, and the one from S_1 to S_2 is called track 2. Slow and fast trains run alternately on these tracks, from S_0 to S_2. The occupation of the tracks by the two train types, in certain time units, is represented as a piece in Figure 10.1. At station S_1, the trains occupy both tracks. The argument for this is that either S_1 has only one platform and the time duration common to both tracks in the figure indicates the dwell time, or there is a sensor at S_1 that signals the passing of a train from track 1 to track 2, and due to the fact that the train has a certain length, both tracks are occupied simultaneously for a certain time. The matrices M, as introduced in Section 1.3, characterizing the shapes of the two pieces are, respectively,

$$M_1 = \begin{pmatrix} 4 & 1 \\ 7 & 4 \end{pmatrix}, \qquad M_2 = \begin{pmatrix} 3 & 2 \\ 4 & 3 \end{pmatrix}. \qquad (10.1)$$

The eigenvalues of these two matrices are $\lambda_1 \stackrel{\text{def}}{=} \lambda(M_1) = 4$ and $\lambda_2 \stackrel{\text{def}}{=} \lambda(M_2) = 3$, respectively. If only slow trains make use of the two tracks, then the capacity of the tracks is determined by λ_1. This can be shown by stacking the corresponding pieces on top of each other, as represented in the left-hand figure in Figure 10.2. Every λ_1 time units a train can pass the tracks. Hence, if travel times are given in minutes, then $60/\lambda_1 = 15$ trains can run on the track per hour. Similarly, every λ_2 time units a fast train can pass the tracks; see the middle figure of Figure 10.2. The right figure in Figure 10.2 represents the case in which slow and fast trains

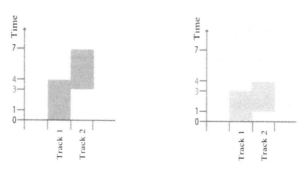

Figure 10.1: The heaps corresponding to the slow (left) and fast train (right).

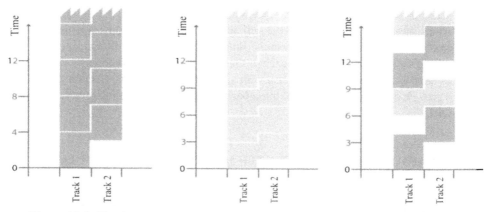

Figure 10.2: The heaps corresponding to a series of slow trains (left), fast trains (middle), and alternating slow and fast trains (right).

alternately use the tracks. The combination of slow and fast trains yields the M matrix

$$M_3 = M_2 \otimes M_1 = \begin{pmatrix} 9 & 6 \\ 10 & 7 \end{pmatrix}.$$

Somewhat surprisingly, perhaps, the eigenvalue of M_3 equals $\lambda_3 = 9$, which is larger than $\lambda_1 \otimes \lambda_2 = 7$. Now, on the average, every $9/2 = 4.5$ time units a train passes along the tracks. An explanation for the phenomenon $\lambda_1 \otimes \lambda_2 < \lambda_3$ is provided by the holes in the heaps of pieces in the right figure in Figure 10.2. One finds the somewhat surprising phenomenon that if a slow train in a long series of slow trains is replaced by a fast train, this causes a decreasing (i.e., slower) performance. Depending on the size of the pieces, one can also have $\lambda_1 \otimes \lambda_2 = \lambda_3$ or even $\lambda_1 \otimes \lambda_2 > \lambda_3$; see exercise 1.

10.2 CAPACITY ASSESSMENT FOR A SERIES OF TUNNELS

The application of this section is concerned with the design of the high-speed connection between Amsterdam and the Belgian/Dutch border as part of the high-speed line between Amsterdam and Paris (known as HSL South). The abbreviation HSL stands for Hoge SnelheidsLijn (Dutch for "high-speed line"). Figure 10.3 shows a map of the planned route of HSL South. The Dutch part of the line will be opera-

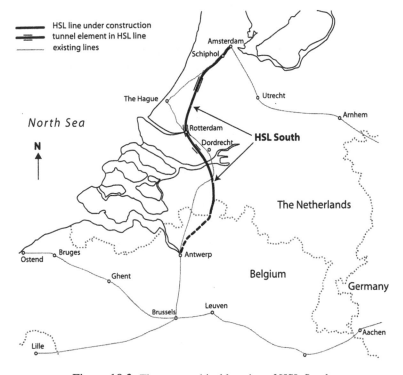

Figure 10.3: The geographical location of HSL South.

tional in 2006, and from then on, the expected traffic load will be eight trains per hour in each direction, increasing up to sixteen trains in each direction per hour by 2015. The line will be used for transporting passengers only (no cargo).

The Dutch part of the line includes three tunnels, each with separated tunnel tubes for both running directions. Figure 10.4 shows the line schematically. From left to right (geographically close to "from north to south"), the tunnels are called Groene Hart, Oude Maas, and Dordtsche Kil, respectively. The corresponding track distances are given in kilometers.

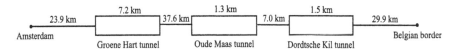

Figure 10.4: Distances along HSL South and its three tunnels.

The Dutch part of HSL South has a total length of approximately 100 kilometers. The maximum permitted speed is 300 km/hour, which is only achieved nearby and inside the Groene Hart tunnel and to the south of the Dordtsche Kil tunnel, towards the Belgian border. Anywhere else, more restrictive speeds apply, in particular at the other two tunnels under consideration (to guarantee that trains will not accelerate too much while descending) and near curves (to prevent derailments). Table 10.1 shows the default travel times for the trains through each of the three tunnels. These run times are based on an accurate description of the HSL South track layout, including, for instance, locations and values for curves, sensors, slopes, the lengths of trains, and (rolling stock–dependent) local speed restrictions.

Tunnel	Direction	Travel time in minutes
Groene	North–South	2.2
Hart	South–North	2.2
Oude	North–South	1.0
Maas	South–North	0.9
Dordtsche	North–South	1.0
Kil	South–North	1.0

Table 10.1: Speed restrictions at HSL South tunnels.

The three tunnels of the Dutch part of HSL South will have no emergency exits to ground level, nor will these three tunnels be equipped with service channels. Instead, each tunnel will be provided with cross-sections connecting both tunnel tubes. This implies that, in case of an emergency (e.g., a train getting stuck in the tunnel while on fire), passengers can only be evacuated via the opposite tunnel tube. Obviously, for this escape route to be safe, the opposite tunnel tube must be free of trains at the time of evacuation.

A special traffic regime has been proposed to guarantee passengers a safe escape route from each of these tunnels at any time. This regime permits the presence of at most one train at a time in each tunnel (thus both tubes). In other words, if a train occupies one of the tunnel tubes, the opposite tube should be immediately blocked for other trains. Consequently, even though HSL South is an entirely double-track line, all three tunnels behave like single-track parts, and as such, the capacity of the entire line may be too restricted in order to establish the intended traffic intensity. An alternative is to build an extra tube (e.g., a service channel). However, drilling a railway-tunnel tube, together with installing the equipment, costs about €50,000 per running meter, which explains the importance of carefully examining the capacity of the line.

10.2.1 Example

A stylized example, representing two tunnel elements in a series, will be given here. For the sake of simplicity, the part between the two tunnel elements is disregarded. We will show that the capacity of the series of tunnels is determined by the minimum of the capacities of each of the tunnels or, equivalently, by the maximum of the two eigenvalues.

Consider Figure 10.5 in which the two tunnel elements are represented as a Petri net. Each tunnel element contains only one token, which indicates that trains cannot pass in opposite directions simultaneously. The travel times for tunnel 1 are 1 time unit for a train from transition q_1 to transition q_2 and 4 time units in the opposite direction. For tunnel 2 the travel times are a and 3, respectively. The max-plus models for each of the tunnels and for the two tunnels together are (the departure time at transition q_i is denoted by x_i)

$$\begin{pmatrix} x_1(k+1) \\ x_2(k+1) \end{pmatrix} = \begin{pmatrix} \varepsilon & 4 \\ \varepsilon & \varepsilon \end{pmatrix} \otimes \begin{pmatrix} x_1(k+1) \\ x_2(k+1) \end{pmatrix} \oplus \begin{pmatrix} \varepsilon & \varepsilon \\ 1 & \varepsilon \end{pmatrix} \otimes \begin{pmatrix} x_1(k) \\ x_2(k) \end{pmatrix},$$

$$\begin{pmatrix} x_2(k+1) \\ x_3(k+1) \end{pmatrix} = \begin{pmatrix} \varepsilon & \varepsilon \\ a & \varepsilon \end{pmatrix} \otimes \begin{pmatrix} x_2(k+1) \\ x_3(k+1) \end{pmatrix} \oplus \begin{pmatrix} \varepsilon & 3 \\ \varepsilon & \varepsilon \end{pmatrix} \otimes \begin{pmatrix} x_2(k) \\ x_3(k) \end{pmatrix},$$

$$\begin{pmatrix} x_1(k+1) \\ x_2(k+1) \\ x_3(k+1) \end{pmatrix} = \begin{pmatrix} \varepsilon & 4 & \varepsilon \\ \varepsilon & \varepsilon & \varepsilon \\ \varepsilon & a & \varepsilon \end{pmatrix} \otimes \begin{pmatrix} x_1(k+1) \\ x_2(k+1) \\ x_3(k+1) \end{pmatrix}$$
$$\oplus \begin{pmatrix} \varepsilon & \varepsilon & \varepsilon \\ 1 & \varepsilon & 3 \\ \varepsilon & \varepsilon & \varepsilon \end{pmatrix} \otimes \begin{pmatrix} x_1(k) \\ x_2(k) \\ x_3(k) \end{pmatrix},$$

from which one immediately obtains

$$\begin{pmatrix} x_1(k+1) \\ x_2(k+1) \end{pmatrix} = \begin{pmatrix} 5 & \varepsilon \\ 1 & \varepsilon \end{pmatrix} \otimes \begin{pmatrix} x_1(k) \\ x_2(k) \end{pmatrix},$$

$$\begin{pmatrix} x_2(k+1) \\ x_3(k+1) \end{pmatrix} = \begin{pmatrix} \varepsilon & 3 \\ \varepsilon & 3+a \end{pmatrix} \otimes \begin{pmatrix} x_2(k) \\ x_3(k) \end{pmatrix},$$

$$\begin{pmatrix} x_1(k+1) \\ x_2(k+1) \\ x_3(k+1) \end{pmatrix} = \begin{pmatrix} 5 & \varepsilon & 7 \\ 1 & \varepsilon & 3 \\ 1+a & \varepsilon & 3+a \end{pmatrix} \otimes \begin{pmatrix} x_1(k) \\ x_2(k) \\ x_3(k) \end{pmatrix}.$$

It is immediately verified that the eigenvalues of these models are 5, $a + 3$, and $\max(5, a + 3, (8 + a)/2) = \max(5, a + 3)$, respectively. Hence, the capacity of both tunnels together is determined by the minimum of the capacities of each of them. This conclusion holds for more general systems than the one given in this example; see exercise 4.

10.2.2 Capacity assessment for the HSL

We now turn to the capacity of the entire HSL. For the sake of simplicity, as in the previous section, we disregard the parts of the line between the tunnel elements. This can be done without any harm, since the double-track segments between the tunnel elements do not determine the capacity of the line. This argument only holds if these tracks can hold an arbitrary number of trains. In practice, a track is divided into safety blocks and only one train is admitted on each block. If such blocks are short compared to the lengths of the tunnel tubes, then the assumption of disregarding the parts of the line between the tunnel elements is a reasonable one.

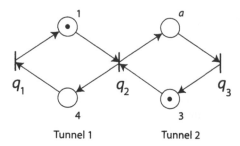

Tunnel 1   Tunnel 2

Figure 10.5: Two tunnel elements in a series represented as a Petri net.

The full mathematical description of the whole line, including the safety blocks and so on, is beyond the scope of this book. Based on the previous section, the capacity for the whole line is here simply defined as the minimum capacity of the three tunnels individually. Table 10.1 then yields that the Groene Hart tunnel is the bottleneck. The eigenvalue equals 4.4 minutes. Hence, in both the north–south and the south–north directions, a train can run every 4.4 minutes. Thus, thirteen trains per hour can run in each direction. Given the current restrictions, the requirement of a traffic load of sixteen trains in each direction (by the year 2015) is not realistic.

10.3 EXERCISES

1. In Section 10.1 two pieces were introduced that were characterized by the two matrices in (10.1). One of the conclusions was that $\lambda(M_2 \otimes M_1) > \lambda(M_1) \otimes \lambda(M_2)$. Show that the same conclusion does not hold for pairs of pieces characterized by

$$M_1 = \begin{pmatrix} 3 & 2 \\ 5 & 4 \end{pmatrix}, \ M_2 = \begin{pmatrix} 3 & 2 \\ 4 & 3 \end{pmatrix}$$

and

$$M_1 = \begin{pmatrix} 4 & 2 \\ 5 & 3 \end{pmatrix}, \ M_2 = \begin{pmatrix} 3 & 2 \\ 5 & 4 \end{pmatrix},$$

respectively.

2. Consider the example of Section 10.2.1 once more but with the difference that each of the places in Figure 10.5 now contains one token. Show that the following holds: the capacity of the two tunnels as a whole equals the minimum of the capacities of each of the tunnels. Can the same conclusion be drawn with an arbitrary number of tokens in each place and with more than two tunnels in the series?

3. An important question not addressed in the definition and determination of the capacity of the three tunnels is whether trains approaching a tunnel element have to wait (or run at slow speed) to enter this element if the train in the opposite direction has not yet left this element. Discuss how the capacity of the whole line might be affected by acceleration and deceleration restrictions on the trains.

4. In the example of Section 10.2.1, the travel times along the four different tracks are 1, 4, a, and 3, respectively. If these travel times are replaced by a, b, c, and d, respectively, then show that the capacity of both tunnels together is still determined

by the minimum of the capacities of each of them (i.e., by the maximum of the two eigenvalues $a + b$ and $c + d$).

10.4 NOTES

A detailed description of a tunnel element is provided in [73], where it is modeled as a closed max-plus system. A study of the HSL South in case of stochastic travel times is given in [58].

The final location of the HSL was the result of long political discussions. One issue was where it should cross the (meandering) border between Belgium and the Netherlands, since each country should, in principle, pay for its part of the line. Whereas the Oude Maas tunnel and the Dordtsche Kil tunnel pass underneath waterways of these names, the Groene Hart (green heart) tunnel is the result of long environmental lobbying so as not to disturb the landscape.

PART III
Extensions

Chapter Eleven

Stochastic Max-Plus Systems

This chapter is devoted to the study of sequences $\{x(k) : k \in \mathbb{N}\}$ satisfying the recurrence relation

$$x(k+1) = A(k) \otimes x(k), \qquad k \geq 0, \tag{11.1}$$

where $x(0) = x_0 \in \mathbb{R}^n_{\max}$ is the initial value and $\{A(k) : k \in \mathbb{N}\}$ is a sequence of $n \times n$ matrices over $\mathbb{R}_{\max}$. In order to develop a meaningful mathematical theory, we need some additional assumptions on $\{A(k) : k \in \mathbb{N}\}$. The approach presented in this chapter assumes that $\{A(k) : k \in \mathbb{N}\}$ is a sequence of random matrices in $\mathbb{R}^{n \times n}_{\max}$, defined on a common probability space. Specifically, we address the case where $\{A(k) : k \in \mathbb{N}\}$ consists of independent identically distributed (i.i.d.) random matrices. The theory is also available for the more general case of $\{A(k) : k \in \mathbb{N}\}$ being an ergodic sequence. However, for ease of exposition, we restrict our presentation to the i.i.d. case.

We focus on the asymptotic growth rate of $x(k)$. Note that $x(k)$ and thus $x(k)/k$ are random variables. We have to be careful about how to interpret the asymptotic growth rate. The key result of this chapter will be that under appropriate conditions the asymptotic growth rate of $x(k)$ defined in (11.1) is, with probability one, a constant.

The stochastic max-plus theory is dissimilar to the deterministic theory developed in this book so far, not only with respect to the applied techniques but also with respect to the obtained results. In the deterministic theory, proofs are usually constructive, and a rich variety of numerical procedures for computing eigenvalues and eigenvectors, for example, can be provided. In the stochastic theory, proofs are usually proofs of existence, and no efficient numerical algorithms for computing, say, the asymptotic growth rate for large-scale models, are available. In highlighting this difference one could say that while deterministic theory comes up with efficient algorithms for computing the asymptotic growth rate, the stochastic theory has to be content with showing that the asymptotic growth rate exists (with probability one) and that it equals some finite constant with probability one. The reader is referred to the notes section for some recently developed numerical approaches.

The stochastic limit theory will be discussed for three different cases of max-plus systems. First, we will study recurrence relations with the properties that (i) the arc set of the communication graph of $A(k)$ is nonrandom and (ii) the communication graph of $A(k)$ is strongly connected with probability one (this is the stochastic equivalent to the study of irreducible matrices). Second, as in deterministic theory, we will drop condition (ii) and study recurrence relations satisfying only condition (i) (this is the stochastic equivalent to the study of reducible matrices). Finally, we will examine recurrence relations not satisfying condition (i).

The chapter is organized as follows. In Section 11.1 basic concepts are introduced for stochastic max-plus recurrence relations (concepts familiar from deterministic theory, such as irreducibility, have to be redefined in a stochastic context). Moreover, examples of stochastic max-plus systems are given. Section 11.2 is devoted to subadditive ergodic theory for stochastic sequences. The limit theory for matrices with property (i) is provided in Section 11.3. Possible relaxations of the rather restrictive conditions needed for the analysis in the latter section are provided in Section 11.4. An overview of the stochastic theory not covered in this book is given in the notes section.

11.1 BASIC DEFINITIONS AND EXAMPLES

For a sequence of square matrices $\{A(k) : k \in \mathbb{N}\}$, we set

$$\bigotimes_{k=l}^{m} A(k) \stackrel{\text{def}}{=} A(m) \otimes A(m-1) \otimes \cdots \otimes A(l+1) \otimes A(l),$$

where $m \geq l$ and $\bigotimes_{k=l}^{m} A(k) \stackrel{\text{def}}{=} E$ otherwise.

A few words on the fundamentals of the stochastic setup are in order here. Let X be a random element in $\mathbb{R}_{\max}$ defined on a probability space $(\Omega, \mathcal{F}, P)$ modeling the underlying randomness.[1] When defining the expected value of X, denoted by $\mathbb{E}[X]$, one has to take care of the fact that X may take value ε $(= -\infty)$ with positive probability. This is reflected in the following extension to $\mathbb{R}_{\max}$ of the usual definition of integrability of a random variable on $\mathbb{R}$. We call $X \in \mathbb{R}_{\max}$ *integrable* if $X \oplus e = \max(X, 0)$ and $X \oplus' e = \min(X, 0)$ are integrable and if $\mathbb{E}[X \oplus e]$ is finite. The expected value of X is then given by $\mathbb{E}[X] = \mathbb{E}[X \oplus e] + \mathbb{E}[X \oplus' e]$. This definition implies that $\mathbb{E}[X] = -\infty$ if $P(X = \varepsilon) > 0$. A random matrix A in $\mathbb{R}_{\max}^{n \times m}$ is called integrable if its elements a_{ij} are integrable for $i \in \underline{n}, j \in \underline{m}$. The expected value of A is given by the matrix $\mathbb{E}[A]$ with elements $\left[\mathbb{E}[A]\right]_{ij} = \mathbb{E}[a_{ij}]$.

In order to define irreducibility for random matrices, we introduce the concept of a *fixed support* of a matrix.

DEFINITION 11.1 *We say that* $\{A(k) : k \in \mathbb{N}\}$ *has fixed support if the set of arcs of the communication graph of* $A(k)$ *is nonrandom and does not depend on* k, *or, more formally, for all* $i, j \in \underline{n}$,

$$\left(\forall k \geq 0 : P\big([A(k)]_{ij} = \varepsilon\big) = 0\right) \vee \left(\forall k \geq 0 : P\big([A(k)]_{ij} = \varepsilon\big) = 1\right).$$

With the definition of fixed support at hand, we say that a random matrix A is *irreducible* if it has fixed support and any sample of A is irreducible with probability one. Hence, for random matrices, irreducibility presupposes fixed support.

Stochasticity occurs quite naturally in real-life railway networks. For example, travel times become stochastic due to, for example, weather conditions or the individual behavior of the driver. Another source of randomness is the time durations

---

[1] It is a assumed that the reader is familiar with basic probability theory.

for boarding or alighting of passengers. Also, the lack of information about the future specification of a railway system, such as the type of rolling stock, the capacity of certain tracks, and so forth, can be modeled by randomness.

Example 11.1.1   *Consider the railway network described in Example 7.2.1 and assume that the travel times are random. More specifically, denote the kth travel time from station S_i to S_{i+1} by $a_{i+1,i}(k)$, for $i \in \underline{2}$ and the kth travel time from station S_3 to S_1 by $a_{1,3}(k)$. It is assumed that the travel times are stochastically independent and that the travel times for a certain track have the same distribution. If we follow the reasoning put forward in Example 7.2.1, together with exercise 5 in Chapter 7, then this system can be modeled through $x(k) = (x_1(k), x_2(k))^\top$, which satisfies*

$$x(k+1) = \begin{pmatrix} a_{21}(k) \oplus a_{13}(k+1) & a_{13}(k+1) \otimes a_{32}(k) \\ a_{21}(k) & a_{32}(k) \end{pmatrix} \otimes x(k),$$

where $x_1(k)$ denotes the kth departure time from station S_1 and $x_2(k)$ denotes the kth departure time from station S_2. Notice that the matrix on the right-hand side of the above equation has fixed support and is irreducible.

Example 11.1.2   *Consider the railway network described in Example 7.3.1, and assume, as in the previous example, that the travel times (and the interarrival times) are stochastically independent and that the travel times for a certain track as well as the interarrival times are identically distributed. Following the reasoning put forward in Example 7.3.1, this system can be modeled through $x(k) = (x_0(k), x_1(k), x_2(k))^\top$, which satisfies*

$$x(k+1) = A(k) \otimes x(k),$$

where the matrix $A(k)$ looks like

$$\begin{pmatrix} a_0(k) & \varepsilon & \varepsilon \\ a_0(k) \otimes a_{10}(k) & e & \varepsilon \\ a_0(k) \otimes a_{10}(k) \otimes a_{21}(k) & a_{21}(k) & e \end{pmatrix},$$

for $k \geq 0$. Observe that $A(k)$ has fixed support but fails to be irreducible.

Example 11.1.3   *Consider a simple railway network consisting of two stations with deterministic travel times between the stations. Specifically, the travel time from Station 2 to Station 1 equals σ', and the dwell time at Station 1 equals d, whereas the travel time from Station 1 to Station 2 equals σ and the dwell time at Station 2 equals d'. At Station 1 there is one platform at which trains can stop, whereas at Station 2 there are two platforms. Three trains circulate in the network. Initially, one train is present at Station 1, one train at Station 2, and the third train is just about to enter Station 2. The time evolution of this network is described by a max-plus linear sequence of vectors $x(k) = (x_1(k), \ldots, x_4(k))^\top$, where $x_1(k)$ is the kth arrival time of a train at Station 1 and $x_2(k)$ is the kth departure time of a train from the Station 1, $x_3(k)$ is the kth arrival time of a train at Station 2, and $x_4(k)$ is the kth departure time of a train from Station 2. Figure 11.1 on the following page shows the Petri net model of this system. The sample-path dynamics*

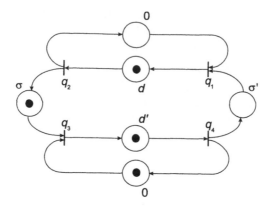

Figure 11.1: The initial state of the railway system with two platforms at Station 2.

of the network with two platforms at Station 2 is given by

$$x_1(k+1) = x_2(k+1) \oplus \big(x_4(k+1) \otimes \sigma'\big),$$
$$x_2(k+1) = x_1(k) \otimes d,$$
$$x_3(k+1) = \big(x_2(k) \otimes \sigma\big) \oplus x_4(k),$$
$$x_4(k+1) = x_3(k) \otimes d',$$

for $k \geq 0$. Replacing $x_2(k+1)$ and $x_4(k+1)$ in the first equation by the expression on the right-hand side of the second and fourth equations above, respectively, yields

$$x_1(k+1) = \big(x_1(k) \otimes d\big) \oplus \big(x_3(k) \otimes d' \otimes \sigma'\big).$$

Hence, for $k \geq 0$,

$$x_1(k+1) = \big(x_1(k) \otimes d\big) \oplus \big(x_3(k) \otimes d' \otimes \sigma'\big),$$
$$x_2(k+1) = x_1(k) \otimes d,$$
$$x_3(k+1) = \big(x_2(k) \otimes \sigma\big) \oplus x_4(k),$$
$$x_4(k+1) = x_3(k) \otimes d',$$

which reads in vector-matrix notation

$$x(k+1) = D_2 \otimes x(k),$$

where

$$D_2 = \begin{pmatrix} d & \varepsilon & d' \otimes \sigma' & \varepsilon \\ d & \varepsilon & \varepsilon & \varepsilon \\ \varepsilon & \sigma & \varepsilon & e \\ \varepsilon & \varepsilon & d' & \varepsilon \end{pmatrix}.$$

Notice that D_2 is irreducible.

Consider the railway network again, but one of the platforms at Station 2 is not available. The initial condition is as in the previous example. Figure 11.2 on the next page shows the Petri net of the system with one blocked platform at Station 2.

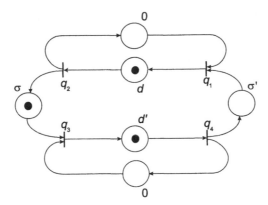

Figure 11.2: The initial state of the railway system with one blocked platform.

Note that the blocking is modeled by the absence of the token in the bottom place, yielding that $x_3(k+1) = (x_2(k) \otimes \sigma) \oplus x_4(k+1)$. Following the line of argument put forward for the network with two platforms at Station 2, one arrives at

$$x(k+1) = D_1 \otimes x(k),$$

where

$$D_1 = \begin{pmatrix} d & \varepsilon & d' \otimes \sigma' & \varepsilon \\ d & \varepsilon & \varepsilon & \varepsilon \\ \varepsilon & \sigma & d' & \varepsilon \\ \varepsilon & \varepsilon & d' & \varepsilon \end{pmatrix}.$$

Notice that D_1 fails to be irreducible.

Assume that whenever a train arrives at Station 2, one platform is blocked with probability p, with $0 < p < 1$. This is modeled by introducing $A(k)$ with distribution

$$P(A(k) = D_1) = p$$

and

$$P(A(k) = D_2) = 1 - p.$$

Then

$$x(k+1) = A(k) \otimes x(k)$$

describes the time evolution of the system with resource restrictions. Notice that $A(k)$ fails to have a fixed support (and that $A(k)$ is thus not irreducible).

11.2 THE SUBADDITIVE ERGODIC THEOREM

Subadditive ergodic theory is based on Kingman's subadditive ergodic theorem and its application to generalized products of random matrices. Kingman's result [56] is formulated in terms of *subadditive processes*. These are double-indexed processes $X = \{X_{ml} : m, l \in \mathbb{N}\}$ satisfying the following conditions:

(S1) For $i, j, k \in \mathbb{N}$, such that $i < j < k$, the inequality $X_{ik} \leq X_{ij} + X_{jk}$ holds with probability one.

(S2) All joint distributions of the process $\{X_{m+1,l+1} : l, m \in \mathbb{N}, l > m\}$ are the same as those of $\{X_{ml} : l, m \in \mathbb{N}, l > m\}$.

(S3) The expected value $g_l = \mathbb{E}[X_{0l}]$ exists and satisfies $g_l \geq -c \times l$ for some finite constant $c > 0$ and all $l \in \mathbb{N}$.

Kingman's celebrated ergodic theorem can now be stated as follows.

THEOREM 11.2 *(Kingman's subadditive ergodic theorem) If $X = \{X_{ml} : m, l \in \mathbb{N}\}$ is a subadditive process, then a finite number ξ exists such that*

$$\xi = \lim_{k \to \infty} \frac{X_{0k}}{k}$$

with probability one and

$$\xi = \lim_{k \to \infty} \frac{\mathbb{E}[X_{0k}]}{k}.$$

The surprising part of Kingman's ergodic theorem is that the random variables X_{0k}/k converge, with probability one, towards the same finite value, which is the limit of $\mathbb{E}[X_{0k}]/k$.

We will apply Kingman's subadditive ergodic theorem to the maximal (resp., minimal) finite element of a matrix. The basic concepts are defined in the following. For $A \in \mathbb{R}_{\max}^{n \times m}$, the minimal finite entry of A, denoted by $||A||_{\min}$, is given by

$$||A||_{\min} = \min\{a_{ij} \mid (i, j) \in \mathcal{D}(A)\},$$

where $||A||_{\min} = \varepsilon' (= +\infty)$ if $\mathcal{D}(A) = \emptyset$. (Recall that $\mathcal{D}(A)$ denotes the set of arcs in the communication graph of A.) In the same vein, we denote the maximal finite entry of $A \in \mathbb{R}_{\max}^{n \times m}$ by $||A||_{\max}$, which implies

$$||A||_{\max} = \max\{a_{ij} \mid (i, j) \in \mathcal{D}(A)\},$$

where $||A||_{\max} = \varepsilon$ if $\mathcal{D}(A) = \emptyset$. A direct consequence of the above definitions is that for any regular $A \in \mathbb{R}_{\max}^{n \times m}$

$$||A||_{\min} \leq ||A||_{\max}.$$

Notice that $||A||_{\min}$ and $||A||_{\max}$ can have negative values. It is easily checked (see exercise 4) that for regular $A \in \mathbb{R}_{\max}^{n \times m}$ and regular $B \in \mathbb{R}_{\max}^{m \times l}$

$$||A \otimes B||_{\max} \leq ||A||_{\max} \otimes ||B||_{\max} \tag{11.2}$$

and

$$||A \otimes B||_{\min} \geq ||A||_{\min} \otimes ||B||_{\min}. \tag{11.3}$$

We now revisit our basic max-plus recurrence relation

$$x(k + 1) = A(k) \otimes x(k),$$

for $k \geq 0$, with $x(0) = x_0$. To indicate the initial value of the sequence, we sometimes use the notation

$$x(k; x_0) = \bigotimes_{l=0}^{k-1} A(l) \otimes x_0, \qquad k \in \mathbb{N}. \tag{11.4}$$

To abbreviate the notation, we set for $m \geq l \geq 0$

$$A[m, l] \stackrel{\text{def}}{=} \bigotimes_{k=l}^{m-1} A(k).$$

With this definition (11.4) can be written as

$$x(k; x_0) = A[k, 0] \otimes x_0,$$

for $k \geq 0$. Notice that for $0 \leq l \leq p \leq m$

$$A[m, l] = A[m, p] \otimes A[p, l]. \tag{11.5}$$

LEMMA 11.3 *Let $\{A(k) : k \in \mathbb{N}\}$ be an i.i.d. sequence of integrable matrices such that $A(k)$ is regular with probability one. Then $\{-\|A[m, l]\|_{\min} : m > l \geq 0\}$ and $\{\|A[m, l]\|_{\max} : m > l \geq 0\}$ are subadditive ergodic processes.*

Proof. For $2 \leq m$ and $0 \leq l < p < m$, we obtain

$$\|A[m, l]\|_{\max} \stackrel{(11.5)}{=} \|A[m, p] \otimes A[p, l]\|_{\max}$$
$$\stackrel{(11.2)}{\leq} \|A[m, p]\|_{\max} + \|A[p, l]\|_{\max},$$

which establishes (S1) for $\|A[m, l]\|_{\max}$. The proof that (S1) also holds for $-\|A[m, l]\|_{\min}$ follows from the same line of argument, where (11.3) is used for establishing the inequality, and the proof is therefore omitted.

The stationarity condition (S2) follows immediately from the i.i.d. assumption for $\{A(k) : k \in \mathbb{N}\}$.

We now turn to condition (S3). The fact that $\{A(k) : k \in \mathbb{N}\}$ is an i.i.d. sequence implies

$$\mathbb{E}\big[\|A[k, 0]\|_{\max}\big] \geq \mathbb{E}\big[\|A[k, 0]\|_{\min}\big]$$
$$\stackrel{(11.3)}{\geq} k \times \mathbb{E}\big[\|A(0)\|_{\min}\big]$$
$$\geq k \times \big(-\big|\mathbb{E}[\|A(0)\|_{\min}]\big|\big).$$

Integrability of $A(0)$ together with regularity implies that $\mathbb{E}[\|A(0)\|_{\min}]$ is finite (for a proof use the fact that $\min(X, Y) \leq |X| + |Y|$). This establishes condition (S3) for $\|A[m, l]\|_{\max}$. For the proof that $-\|A[m, l]\|_{\min}$ satisfies (S3), notice that (11.2) implies

$$-\mathbb{E}\big[\|A[k, 0]\|_{\min}\big] \geq -\mathbb{E}\big[\|A[k, 0]\|_{\max}\big] \geq -k \times \mathbb{E}\big[\|A(0)\|_{\max}\big].$$

Since integrability of $A(0)$ together with regularity implies that $\mathbb{E}\big[\|A(0)\|_{\max}\big]$ is finite, we have proved the claim. □

The above lemma shows that Kingman's subadditive ergodic theorem can be applied to $\|A[k, 0]\|_{\min}$ and $\|A[k, 0]\|_{\max}$. The precise statement is given in the following theorem.

THEOREM 11.4 *Let* $\{A(k) : k \in \mathbb{N}\}$ *be an i.i.d. sequence of integrable matrices such that* $A(k)$ *is regular with probability one. Then, finite constants* λ^{top} *and* λ^{bot} *exist such that with probability one*

$$\lambda^{\text{bot}} \overset{\text{def}}{=} \lim_{k \to \infty} \frac{1}{k} \|A[k, 0]\|_{\min} \leq \lambda^{\text{top}} \overset{\text{def}}{=} \lim_{k \to \infty} \frac{1}{k} \|A[k, 0]\|_{\max}$$

and

$$\lambda^{\text{bot}} = \lim_{k \to \infty} \frac{1}{k} \mathbb{E}\Big[\|A[k, 0]\|_{\min} \Big], \qquad \lambda^{\text{top}} = \lim_{k \to \infty} \frac{1}{k} \mathbb{E}\Big[\|A[k, 0]\|_{\max} \Big].$$

The constant λ^{top} is called the *top* or *maximal Lyapunov exponent* of $\{A(k) : k \in \mathbb{N}\}$, and λ^{bot} is called the *bottom* or *minimal Lyapunov exponent* of $\{A(k) : k \in \mathbb{N}\}$. The top and bottom Lyapunov exponents of $A(k)$ are related to the asymptotic growth rate of $x(k)$ defined in (11.1) as follows. The top Lyapunov exponent equals the asymptotic growth rate of the maximal entry of $x(k)$, and the bottom Lyapunov exponent equals the asymptotic growth rate of the minimal entry of $x(k)$. The precise statement is given in the following corollary.

COROLLARY 11.5 *Let* $\{A(k) : k \in \mathbb{N}\}$ *be an i.i.d. sequence of integrable matrices such that* $A(k)$ *is regular with probability one. Then, for any finite and integrable initial condition* x_0, *it holds with probability one that*

$$\lambda^{\text{bot}} = \lim_{k \to \infty} \frac{\|x(k; x_0)\|_{\min}}{k} \leq \lambda^{\text{top}} = \lim_{k \to \infty} \frac{\|x(k; x_0)\|_{\max}}{k}$$

and

$$\lambda^{\text{bot}} = \lim_{k \to \infty} \frac{1}{k} \mathbb{E}\Big[\|x(k; x_0)\|_{\min} \Big], \qquad \lambda^{\text{top}} = \lim_{k \to \infty} \frac{1}{k} \mathbb{E}\Big[\|x(k; x_0)\|_{\max} \Big].$$

Proof. Note that $x(k; x_0) = A[k, 0] \otimes x_0$ for any $k \in \mathbb{N}$. Provided that x_0 is finite, it is easily checked (see exercise 4) that

$$\|A[k, 0]\|_{\min} \otimes \|x_0\|_{\min} \leq \|x(k; x_0)\|_{\min} \leq \|A[k, 0]\|_{\min} \otimes \|x_0\|_{\max}.$$

Dividing the above row of inequalities by k and letting k tend to ∞ yields

$$\lim_{k \to \infty} \frac{1}{k} \|x(k; x_0)\|_{\min} = \lambda^{\text{bot}}$$

with probability one. The proof for the other limit follows from the same line of argument.

The arguments used for the proof of the first part of the corollary remain valid when expected values are applied (we omit the details). This concludes the proof of the corollary. $\square$

A sufficient condition for $A(k)$ to be regular with probability one is the irreducibility of $A(k)$. Therefore, in the literature, Theorem 11.4 and Corollary 11.5 are often stated with irreducibility (instead of regularity) as a condition.

11.3 MATRICES WITH FIXED SUPPORT

11.3.1 Irreducible matrices

In this section, we consider i.i.d. sequences $\{A(k) : k \in \mathbb{N}\}$ of integrable and irreducible matrices such that with probability one finite entries are bounded from below by a finite constant. As we will show in the following theorem, the setting of this section implies that $\lambda^{\text{top}} = \lambda^{\text{bot}}$, which in particular implies convergence of $x_i(k)/k$ as k tends to ∞, for $i \in \underline{n}$. The main technical result is provided in the following lemma.

LEMMA 11.6 *Let $D \in \mathbb{R}_{\max}^{n \times n}$ be a nonrandom irreducible matrix such that its communication graph has cyclicity σ. If $A(k) \geq D$ with probability one, for any k, then integers L and N exist such that for any $k \geq N$*

$$\|x(k)\|_{\min} \geq \|x(k - L)\|_{\max} + (\|D^{\otimes \sigma}\|_{\min})^{\otimes L}.$$

Proof. Denote the communication graph of D by $\mathcal{G} = (\mathcal{N}, \mathcal{D})$, and let $\mathcal{G}$ be of cyclicity one. Denote the number of elementary circuits in $\mathcal{G}$ by q, and let β_i denote the length of circuit ξ_i, for $i \in \underline{q}$. Then the greatest common divisor of $\{\beta_1, \ldots, \beta_q\}$ is equal to one. According to Theorem 3.2 a natural number N exists such that for all $\kappa \geq N$ there are integers $n_1, \ldots, n_q \geq 0$ such that $\kappa = n_1\beta_1 + \cdots + n_q\beta_q$.

Let l_{ij} denote the minimal length of a path from j to i containing *all* nodes of $\mathcal{G}$. Such paths exist because D is irreducible (and, hence, $\mathcal{G}$ is strongly connected). Let the maximal length of all these paths be denoted by l, i.e., $l = \max_{i,j \in \underline{n}} l_{ij}$.

Next, choose an L with $L \geq N + l$. Then for any $i, j \in \underline{n}$, there is a path from j to i of length L. Indeed, take any $i, j \in \underline{n}$ and choose a path, as mentioned above, from j to i containing *all* nodes of $\mathcal{G}$ and having minimal length l_{ij}. Clearly, the path has at least one node in common with each of the q circuits in $\mathcal{G}$. As $L - l_{ij} \geq N$, there are integers $n_1, \ldots, n_q \geq 0$ such that $L - l_{ij} = n_1\beta_1 + \cdots + n_q\beta_q$. Hence, by adding n_1 copies of circuit ξ_1, and so on, up to n_q copies of circuit ξ_q to the chosen path from i to j of length l_{ij}, a new path from j to i is created of length L.

In graph-theoretical terms, the element $[A(k, k - L)]_{ij}$ denotes the maximal weight of a path of length L from node j to node i on the "interval" $[k - L, k)$. Since $A[k, k - L] \geq D^{\otimes L}$ by assumption, it follows that for all $k \geq N$ and all $i \in \underline{n}$

$$x_i(k) = \bigoplus_{j=1}^{n} [A(k, k - L)]_{ij} \otimes x_j(k - L)$$

$$\geq \bigoplus_{j=1}^{n} \left[D^{\otimes L} \right]_{ij} \otimes x_j(k - L)$$

$$\geq \bigoplus_{j=1}^{n} (\|D\|_{\min})^{\otimes L} \otimes x_j(k - L)$$

$$\geq (\|D\|_{\min})^{\otimes L} \otimes \bigoplus_{j=1}^{n} x_j(k - L),$$

implying that

$$\|x(k)\|_{\min} \geq \|x(k - L)\|_{\max} + (\|D\|_{\min})^{\otimes L}, \qquad \forall k \geq N.$$

Similarly as in Lemma 3.3 it can be shown that if $\mathcal{G}(D)$ has cyclicity σ, then $\mathcal{G}(D^{\otimes\sigma})$ has cyclicity one. Applying the arguments put forward above to $D^{\otimes\sigma}$ extends the result to the case of matrices with cyclicity greater than one. $\qquad\square$

The condition that $A(k) \geq D$ with probability one for any $k \in \mathbb{N}$ and with D being irreducible will be referred to as condition (H_1).

(H_1) *There exists a nonrandom irreducible matrix D such that $A(k) \geq D$ for any $k \in \mathbb{N}$, with probability one.*

Notice that Example 11.1.1 satisfies (H_1), whereas Example 11.1.2 and Example 11.1.3 fail to satisfy (H_1). Lemma 11.6 provides the main technical means for establishing sufficient conditions for equality of maximal, minimal, and individual growth rates. The precise statement is provided in the following theorem.

THEOREM 11.7 *Let $\{A(k): k \in \mathbb{N}\}$ be a random sequence of integrable matrices satisfying (H_1). For $x(k)$ defined in (11.1) it holds, with probability one, that*

$$\lim_{k\to\infty} \frac{1}{k}\|x(k; x_0)\|_{\min} = \lim_{k\to\infty} \frac{1}{k}x_i(k; x_0) = \lim_{k\to\infty} \frac{1}{k}\|x(k; x_0)\|_{\max}$$

for any $i \in \underline{n}$ and any finite initial state x_0.

Proof. Let D be given as in (H_1); then D satisfies the condition put forward in Lemma 11.6, and finite positive numbers L and N exist such that for $k \geq N$

$$\|x(k; x_0)\|_{\min} \geq \|x(k - L; x_0)\|_{\max} + (\|D^{\otimes\sigma}\|_{\min})^{\otimes L},$$

where σ denotes the cyclicity of the communication graph of D. Dividing both sides of the above inequality by k and letting k tend to ∞ yields

$$\lim_{k\to\infty} \frac{1}{k}\|x(k; x_0)\|_{\min} \geq \lim_{k\to\infty} \frac{1}{k}\|x(k; x_0)\|_{\max}, \qquad (11.6)$$

for any finite initial vector x_0. The existence of the above limits is guaranteed by Corollary 11.5, where we use the fact that (H_1) implies that $A(k)$ is regular with probability one. Following the line of argument in the proof of Corollary 11.5, the limits in (11.6) are independent of the initial state.

Combining (11.6) with the obvious fact that $\|x(k; x_0)\|_{\max} \geq x_j(k; x_0) \geq \|x(k; x_0)\|_{\min}$, for $j \in \underline{n}$, proves the claim. $\qquad\square$

By Theorem 11.7, integrability of $A(k)$ together with (H_1) is a sufficient condition for the top and bottom Lyapunov exponent to coincide. Moreover, a random matrix $A(k)$ satisfies condition (H_1) if $A(k)$ is irreducible and if, with probability one, all finite elements are bounded from below by a finite number. Combining this with Theorem 11.4 and Corollary 11.5, we arrive at the following limit theorem for i.i.d. sequences of irreducible matrices.

THEOREM 11.8 *Let $\{A(k) : k \in \mathbb{N}\}$ be an i.i.d. sequence of integrable and irreducible matrices such that with probability one all finite elements are bounded* *from below by a finite number. Then, it holds that* $\lambda \stackrel{\text{def}}{=} \lambda^{\text{top}} = \lambda^{\text{bot}}$, *and with probability one for all* $i, j \in \underline{n}$

$$\lim_{k \to \infty} \frac{1}{k} [A[k, 0]]_{ij} = \lim_{k \to \infty} \frac{1}{k} \mathbb{E}\Big[[A[k, 0]]_{ij} \Big] = \lambda.$$

Moreover, for any finite integrable initial condition x_0 *it holds with probability one that*

$$\lim_{k \to \infty} \frac{x_j(k; x_0)}{k} = \lim_{k \to \infty} \frac{1}{k} \mathbb{E}\Big[x_j(k; x_0) \Big] = \lambda, \qquad j \in \underline{n}.$$

The constant λ, defined in Theorem 11.8, is referred to as the *max-plus Lyapunov exponent* of the sequence of random matrices $\{A(k) : k \in \mathbb{N}\}$. There is no ambiguity in denoting the Lyapunov exponent of $\{A(k) : k \in \mathbb{N}\}$ and the eigenvalue of a matrix A by the same symbol, since the Lyapunov exponent of $\{A(k) : k \in \mathbb{N}\}$ is just the eigenvalue of A whenever $A(k) = A$ for all $k \in \mathbb{N}$. To see this, compare Theorem 11.8 with Lemma 3.12.

The system in Example 11.1.1 satisfies the conditions in Theorem 11.8, and the existence of the Lyapunov exponent is thus guaranteed. Notice that the systems in Examples 11.1.2 and 11.1.3 cannot be analyzed by Theorem 11.8.

11.3.2 Reducible matrices

Now suppose that $A(k)$ has a fixed support and drop the assumption that it is irreducible. To deal with reducible matrices $A(k)$, we decompose $A(k)$ into its irreducible parts. The limit theorem, to be presented shortly, then states that the Lyapunov exponent of the overall matrix equals the maximum of the Lyapunov exponent of its irreducible components. This result presents the stochastic version of Theorem 3.17.

Let $\{A(k) : k \in \mathbb{N}\}$ be a sequence of matrices in $\mathbb{R}^{n \times n}_{\max}$ with fixed support, and consider the associated communication graph of $A(k)$ (with nonrandom arc set). For $i \in \underline{n}$, $[i]$ denotes the set of nodes of the m.s.c.s. that contains node i, and denote by $\lambda_{[i]}$ the Lyapunov exponent associated to the matrix obtained by restricting $A(k)$ to the nodes in $[i]$. We state the theorem without proof. A proof can, for example, be found in [5].

THEOREM 11.9 *Let* $\{A(k) : k \in \mathbb{N}\}$ *be an i.i.d. sequence of integrable matrices in* $\mathbb{R}^{n \times n}_{\max}$ *with fixed support such that with probability one all finite elements are bounded from below by a finite number. For any finite integrable initial value* x_0, *it holds with probability one that*

$$\lim_{k \to \infty} \frac{x_j(k; x_0)}{k} = \lim_{k \to \infty} \frac{1}{k} \mathbb{E}\Big[x_j(k; x_0) \Big] = \lambda_j,$$

with

$$\lambda_j = \bigoplus_{i \in \pi^*(j)} \lambda_{[i]}, \qquad j \in \underline{n}.$$

The system in Example 11.1.2 satisfies the conditions in Theorem 11.9, and the existence of the Lyapunov exponent is thus guaranteed. Notice that the system in Example 11.1.3 cannot be analyzed by Theorem 11.9 because its support is not fixed.

11.4 BEYOND FIXED SUPPORT

In this section we discuss possible relaxations of the conditions put forward in Theorem 11.8. The main technical condition is the following.

(H$_2$) *There exists a nonrandom irreducible matrix D such that*
$$P(A(k) \geq D) \geq p, \quad k \in \mathbb{N},$$
for some $p \in (0, 1]$.

Condition (H$_2$) suffices to guarantee that the top and bottom Lyapunov exponent coincide. The precise statement is given in the following lemma.

LEMMA 11.10 *Let $\{A(k) : k \in \mathbb{N}\}$ be an i.i.d. sequence of integrable matrices such that $A(0)$ is regular with probability one. If condition (H$_2$) holds, then the top and bottom Lyapunov exponents of $\{A(k) : k \in \mathbb{N}\}$ coincide.*

Proof. By Lemma 11.6, there exists an integer L such that there is a path of length L from any node j to any node i in the graph of D with weight at least $(\|D\|_{\min})^{\otimes L}$, where we assume, for ease of exposition, that the communication graph of D is of cyclicity one. Consider the event that for some k it holds that
$$\forall l \in \underline{L} : \quad A(k - l) \geq D. \tag{11.7}$$
On this event,
$$\bigotimes_{l=1}^{L} A(k - l) \geq D^{\otimes L},$$
and in accordance with Lemma 11.6 it follows that
$$\|x(k)\|_{\min} \geq \|x(k - L)\|_{\max} + (\|D\|_{\min})^{\otimes L}. \tag{11.8}$$
Notice that by assumption (H$_2$) the event characterized in (11.7) occurs at least with probability $p^L > 0$. Let $\{\tau_m\}$ be the sequence of times k when the event characterized in (11.7) occurs. The i.i.d. assumption implies that $\tau_m < \infty$ for $m \in \mathbb{N}$ and that $\lim_{m \to \infty} \tau_m = \infty$. By inequality (11.8),
$$\|x(\tau_m)\|_{\min} \geq \|x(\tau_m - L)\|_{\max} + (\|D\|_{\min})^{\otimes L},$$
and dividing both sides of the above inequality by τ_m and letting m tend to ∞ yields with probability one
$$\lim_{m \to \infty} \frac{1}{\tau_m} \|x(\tau_m)\|_{\min} \geq \lim_{m \to \infty} \frac{1}{\tau_m} \|x(\tau_m)\|_{\max}.$$
The existence of the top and the bottom Lyapunov exponents is guaranteed by Corollary 11.5, and the above inequality for a subsequence of $x(k)$ is sufficient to establish equality of the top and bottom Lyapunov exponents. $\square$

Lemma 11.10 allows us to extend Theorem 11.8 to matrices that fail to have a fixed support. More precisely, the fixed support condition can be replaced by the assumption that $A(k)$ is, with positive probability, bounded from below by an irreducible nonrandom matrix. Notice that D_2 in Example 11.1.3 is irreducible, and $\{A(k) : k \in \mathbb{N}\}$ in Example 11.1.3 thus satisfied condition (H$_2$) (take $D = D_2$). The extended version of Theorem 11.8 thus applies to this example.

11.5 EXERCISES

1. Show that if $A \in \mathbb{R}_{\max}^{n \times m}$ and $B \in \mathbb{R}_{\max}^{m \times l}$ are integrable, then $A \otimes B$ is integrable.

2. Show that if $A \in \mathbb{R}_{\max}^{n \times m}$ and $B \in \mathbb{R}_{\max}^{m \times l}$ are regular with probability one, then $A \otimes B$ is regular with probability one.

3. Show that if A is regular with probability one, then $\|A\|_{\min}$ and $\|A\|_{\max}$ are finite with probability one.

4. Let $A \in \mathbb{R}_{\max}^{n \times m}$ and $B \in \mathbb{R}_{\max}^{m \times l}$ be regular. Show that

$$\|A\|_{\min} \otimes \|B\|_{\min} \le \|A \otimes B\|_{\min}, \qquad \|A \otimes B\|_{\max} \le \|A\|_{\max} \otimes \|B\|_{\max},$$

and

$$\|A\|_{\min} \otimes \|B\|_{\min} \le \|A \otimes B\|_{\min} \le \|A\|_{\min} \otimes \|B\|_{\max}.$$

5. Suppose that for $\{x(k) : k \in \mathbb{N}\}$ defined in (11.1) it holds that $\mathbb{E}[x(k+1) - x(k)]$ converges to $\mathbf{u}[\lambda]$ as k tends to ∞ for some finite constant λ. Show that this implies that λ is the Lyapunov exponent of $\{A(k) : k \in \mathbb{N}\}$. (Hint: Use a Cesaro averaging argument.)

6. Show that condition (H_2) can be relaxed as follows. There exists a finite number M and nonrandom matrices $D_i \in \mathbb{R}_{\max}^{n \times n}$, for $i \in \underline{M}$, such that $D_M \otimes \cdots \otimes D_2 \otimes D_1$ is irreducible and $P(A(k) \ge D_i) > 0$, for $i \in \underline{M}$.

7. Consider the system $x(k+1) = A(k) \otimes x(k)$, with $A(k) = D_1$ with probability 0.5 and $A(k) = D_2$, also with probability 0.5. The matrices D_1 and D_2 are taken from Example 11.1.3 into which the numerical values $\sigma = \sigma' = d = 1$ and $d' = 2$ are substituted. The elements in the sequence $A(k)$, $k \in \mathbb{N}$, are assumed to be independent.

   - If one starts with an arbitrary initial state, say, $x(0) = (0,0,0,0)^{\top}$, then one considers the evolution of the state $\overline{x(k)}$ in the projective space (see Section 1.4). For $\overline{x(1)}$ one gets two possibilities according to whether D_1 or D_2 was the transition matrix. Each of these possibilities leads to two possible $\overline{x(2)}$ states and so on. Show that this projective space consists of ten elements and that the set of absorbing states consists of $\overline{x}^{(1)} \overset{\text{def}}{=} \overline{(0,0,-1,-1)^{\top}}$, $\overline{x}^{(2)} \overset{\text{def}}{=} \overline{(0,-1,-1,-1)^{\top}}$, and $\overline{x}^{(3)} \overset{\text{def}}{=} \overline{(0,-1,-2,-1)^{\top}}$.

   - A Markov chain can be constructed with these three states, as indicated in Figure 11.3, left.

Figure 11.3: Markov chain with transition probabilities (left) and with time durations (right).

Show that the stationary distribution for this Markov chain is $p_1 = p_3 = 0.25$ and $p_2 = 0.5$, where p_i corresponds to $\overline{x}^{(i)}$.

- The Lyapunov exponent can be calculated as

$$\lambda = p_1 t_{21} + p_2 \left(\frac{1}{2} t_{22} + \frac{1}{2} t_{32} \right) + p_3 t_{13} = \frac{7}{4},$$

where the t_{ij}'s are the time durations as indicated in Figure 11.3, right.

- Note that $\lambda(D_1) = 2$ and $\lambda(D_2) = \frac{5}{3}$ and that $\frac{1}{2}(\lambda(D_1) + \lambda(D_2)) \neq \frac{7}{4}$.

8. Show that condition (H_2) in Lemma 11.10 can be replaced by the following (weaker) condition:

(H_3) A nonrandom irreducible matrix D and a fixed number N exist such that

$$P \left(\bigotimes_{i=k+1}^{k+N} a(i) \geq D \right) \geq p$$

for some $p \in (0, 1]$.

11.6 NOTES

Example 11.1.3 is an adaptation of an example by Baccelli and Hong [6]. A different approach to stability theory elaborating on the projective space can be found in [63].

A discussion of max-plus linearity in terms of queueing systems can be found in [51]. A max-plus-based analysis of a train network with stochastic travel times can be found in [58]. In [52], a control-theoretic approach to train networks with stochastic travel times based on a max-plus model can be found.

Computing the Lyapunov exponent exactly is a long-standing problem. Upper and lower bounds can be found in [7] and [8]. Approaches that use parallel simulation to estimate the growth rate $x_j(k)/k$ for large k are described in [4]. A classical reference on Lyapunov exponents of products of random matrices is [14], and a more recent one, dedicated to non-negative matrices, is [53].

Based on a limit theorem for Markov chains, strong limit theorems for max-plus systems providing results on Lyapunov exponents have been developed; see [70], [74] , [79], and [84]. Exercise 7 is an example of this approach, where the Lyapunov exponent can actually be computed. Unfortunately, apart from simple problems, computing the Lyapunov exponent in this manner becomes extremely difficult.

The lack of numerical approaches for stochastic max-plus systems has lead to an increased interest in Taylor series approximations of performance characteristics of max-plus systems. The pioneering paper of Baccelli and Schmidt [10] has initiated an ongoing search for better and more efficient algorithms for approximately computing characteristics of stochastic max-plus systems. Recent results in this area are [3], [6], and [39].

One of the celebrated results in the field of stochastic max-plus theory is the extension of Loyne's result on the stability of waiting times in the G/G/1 queue [60] to max-plus linear queueing systems. Readers interested in the max-plus theory of waiting times are referred to [5], [9], and [63].

In [13] the stochastic approach of this chapter is combined with the so-called model predictive control problem, which is well known in system theory.

Chapter Twelve

Min-Max-Plus Systems and Beyond

In this chapter min-max-plus systems will be studied. Such systems can be viewed as an extension of max-plus systems in the sense that in addition to the max and plus operators, the min(imization) operator is now also allowed. This gives more flexibility with respect to modeling issues. At the end of this chapter, we will briefly discuss the imbedding of min-max-plus systems in the even more general class of nonexpansive systems.

12.1 MIN-MAX-PLUS SYSTEMS

12.1.1 Introduction and classification

Min-max-plus systems are described by expressions in which the three operations minimization, maximization, and addition appear. They can be viewed as an extension of max-plus expressions in the sense that minimization has been added as a possible operation. For instance,

$$\min\left(x_1 + 3, \max\left(x_2 - 2, \min(x_1 + 7, x_3)\right), \max(x_3 + 1, x_4 + 2)\right), \quad (12.1)$$

or, equivalently in the min-max-plus notation,

$$(x_1 \otimes 3) \oplus' \left((x_2 \otimes -2) \oplus ((x_1 \otimes 7) \oplus' x_3)\right) \oplus' \left((x_3 \otimes 1) \oplus (x_4 \otimes 2)\right),$$

is a min-max-plus expression. It will be clear that the class of min-max-plus systems is richer than the class of max-plus systems; that is, one can describe more general phenomena in the former class.

Example 12.1.1 *Think of the preparation of different meals, each one consisting of various dishes, to be served at the same time in a restaurant. For the preparation, one needs the ingredients at the right time. The preparation will furthermore depend on the labor involved, such as washing the lettuce. The earliest time instant at which all meals can be served is after the last time instant at which all dishes are ready. In the process of preparation, some dishes will probably already be available before the time of serving. Depending on the particulars of these dishes, one should prepare them as late as possible so as not to ruin their taste during the idle time between being ready and being served. Hence, this idle time must be kept to a minimum. Thus, the cook faces a decision process with the maximization operator (the maximum of all time instants at which all dishes are ready), the minimum operator (minimizing the idle times), and the addition (the time needed for washing the ingredients for the salad, boiling the water, etc.).*

Formally, min-max-plus systems can be introduced by means of a recursive definition scheme.

DEFINITION 12.1 *A min-max-plus expression is an expression that can be thought of as being generated by the following scheme. Variables $x_1, x_2, \ldots, x_n$ taking values in $\mathbb{R}$ are min-max-plus expressions. If f is a min-max-plus expression, then $f \otimes a$ is a min-max-plus expression, where $a \in \mathbb{R}$ is a parameter. If, in addition, g is a min-max-plus expression, then $f \oplus' g$ and $f \oplus g$ are min-max-plus expressions. No other expressions are min-max-plus expressions.*

The most elementary min-max-plus expression is simply a variable, like x_i or x_j. One can add constants to these variables $x_i \otimes a$ and take the minimum $x_i \oplus' x_j$ or take the maximum $x_i \oplus x_j$. These latter expressions can be combined once more by means of the $\oplus'$ or $\oplus$ operators to obtain more complex expressions. In this way one can continue. It is easily seen that (12.1) is indeed a min-max-plus expression; however, neither $(x_1 \otimes x_2) \oplus (x_3 \otimes -1)$ nor $x_1 \oplus' 4$ are.

By means of the identities

$$a \oplus (b \oplus c) = a \oplus b \oplus c, \tag{12.2}$$

$$a \oplus' (b \oplus' c) = a \oplus' b \oplus' c, \tag{12.3}$$

$$c \oplus (a \oplus' b) = (c \oplus a) \oplus' (c \oplus b), \tag{12.4}$$

$$c \oplus' (a \oplus b) = (c \oplus' a) \oplus (c \oplus' b), \tag{12.5}$$

each min-max-plus expression f can be transformed into the *conjunctive normal form*; that is, we have

$$f = f_1 \oplus' f_2 \oplus' \cdots \oplus' f_p,$$

for some finite $p \in \mathbb{N}$ and where each f_i is a max-plus expression, i.e.,

$$f_i = (x_1 \otimes a_{i1}) \oplus (x_2 \otimes a_{i2}) \oplus \cdots \oplus (x_n \otimes a_{in}),$$

with $a_{ij} \in \mathbb{R}_{\max}$. The adjective *conjunctive* is related to the logical *and*, which is mathematically often written as $\bigwedge$. The latter symbol refers to the minimum operator. Hence, we have the name *conjunctive normal form*. Each min-max-plus expression f can equally well be transformed into the *disjunctive normal form*; that is, we have

$$f = f_1 \oplus f_2 \oplus \cdots \oplus f_q$$

for some finite $q \in \mathbb{N}$ and where now each f_i is a min-plus expression, i.e.,

$$f_i = (x_1 \otimes b_{i1}) \oplus' (x_2 \otimes b_{i2}) \oplus' \cdots \oplus' (x_n \otimes b_{in}),$$

with $b_{ij} \in \mathbb{R}_{\min}$, defined in Example 1.1.1. The adjective *disjunctive* is related to the logical *or*, which in mathematical expressions becomes the maximum operator, often written as $\bigvee$.

Example 12.1.2 *Consider expression (12.1):*

$$(x_1 \otimes 3) \oplus' \big((x_2 \otimes -2) \oplus ((x_1 \otimes 7) \oplus' x_3)\big) \oplus' \big((x_3 \otimes 1) \oplus (x_4 \otimes 2)\big)$$

$$\overset{(12.4)}{=} (x_1 \otimes 3) \oplus' \big(((x_2 \otimes -2) \oplus (x_1 \otimes 7)) \oplus' ((x_2 \otimes -2) \oplus x_3)\big)$$
$$\oplus' \big((x_3 \otimes 1) \oplus (x_4 \otimes 2)\big)$$

$$\overset{(12.3)}{=} (x_1 \otimes 3) \oplus' \big((x_2 \otimes -2) \oplus (x_1 \otimes 7)\big) \oplus' \big((x_2 \otimes -2) \oplus (x_3)\big)$$
$$\oplus' \big((x_3 \otimes 1) \oplus (x_4 \otimes 2)\big).$$

The last expression is in the conjunctive normal form.

An expression in conjunctive normal form can also be written in disjunctive normal form (or the other way around), as is shown by the following example.

Example 12.1.3 *We have*

$$(c \oplus d) \oplus' (a \oplus b) \overset{(12.5)}{=} (c \oplus' (a \oplus b)) \oplus (d \oplus' (a \oplus b))$$
$$\overset{(12.5)}{=} ((c \oplus' a) \oplus (c \oplus' b)) \oplus ((d \oplus' a) \oplus (d \oplus' b))$$
$$\overset{(12.2)}{=} (c \oplus' a) \oplus (c \oplus' b) \oplus (d \oplus' a) \oplus (d \oplus' b).$$

DEFINITION 12.2 *A min-max-plus function of dimension n is a mapping $\mathcal{M}$: $\mathbb{R}^n \to \mathbb{R}^n$, where the components $\mathcal{M}_i$ of $\mathcal{M}$ are min-max-plus expressions of the n variables $x_1, x_2, \ldots, x_n$.*

Max-plus algebra and min-plus algebra have been already introduced in Section 0.5 and formally defined in Section 1.1. Please note that the min operator is a nonlinear operator in max-plus algebra and that the max operator is nonlinear in min-plus algebra. The following are properties of $\mathcal{M}$:

- $\mathcal{M}$ is monotone; that is, if $x, \bar{x} \in \mathbb{R}^n$ such that $x \leq \bar{x}$, then $\mathcal{M}(x) \leq \mathcal{M}(\bar{x})$, where these inequalities must be interpreted componentwise.

- $\mathcal{M}$ is homogeneous; that is, $\mathcal{M}(\alpha \otimes x) = \alpha \otimes \mathcal{M}(x)$ for any scalar $\alpha \in \mathbb{R}$ and any $x \in \mathbb{R}^n$, where the scalar multiplication in both cases refers to componentwise addition of α.

- $\mathcal{M}$ is nonexpansive; that is, $||\mathcal{M}(x) - \mathcal{M}(\bar{x})||_\infty \leq ||x - \bar{x}||_\infty$ for arbitrary $x, \bar{x} \in \mathbb{R}^n$, where $||.||_\infty$ refers to the supremum norm (i.e., the l^∞-norm; see Section 3.2). For a further discussion of nonexpansive mappings, see Section 12.2.2.

In the scientific literature, functions that satisfy the above three properties are called *topical functions*. The class of topical functions is essentially larger than the class of min-max-plus functions. See the notes section of this chapter for some further information and also Example 12.2.1.

If all components $\mathcal{M}_i(x)$ are (re)written in the conjunctive normal form, then we can formally write $\mathcal{M}(x) = \min_{j \in J}(A_j \otimes x)$, where J is a finite set and where all A_j are matrices over $\mathbb{R}_{\max}$ with size $n \times n$. Such a representation is called a *max-representation* of $\mathcal{M}$. If $\mathcal{M}$ is (re)written as $\max_{j \in J'}(B_j \otimes' x)$, with J' being a finite set and the B_j all being $n \times n$ matrices over $\mathbb{R}_{\min}$, then the latter representation of $\mathcal{M}$ is called a *min-representation*. The max-representation (and similarly the min-representation) of a mapping $\mathcal{M}$ is not necessarily unique, as is shown by the next example.

Example 12.1.4 *If*

$$\mathcal{M}_1(x) = \min\left(\max(x_1 + 1, x_2 + 4), x_2\right),$$
$$\mathcal{M}_2(x) = \min\left(\max(x_1 + 3, x_2 + 2), \max(x_1 + 5, x_2 - 2)\right),$$

with $x = (x_1, x_2)^\top$, *then both* $\mathcal{M}(x) = \min_{j=1,2}(A_j \otimes x)$ *and* $\mathcal{M}(x) = \min_{j=3,4}(A_j \otimes x)$ *are max-representations, where*

$$A_1 = \begin{pmatrix} 1 & 4 \\ 3 & 2 \end{pmatrix}, A_2 = \begin{pmatrix} \varepsilon & 0 \\ 5 & -2 \end{pmatrix}, A_3 = \begin{pmatrix} 1 & 4 \\ 5 & -2 \end{pmatrix}, A_4 = \begin{pmatrix} \varepsilon & 0 \\ 3 & 2 \end{pmatrix}.$$

DEFINITION 12.3 *A min-max-plus system of dimension n is a system with state $x(k) = (x_1(k), x_2(k), \ldots, x_n(k))^\top$, which evolves according to $x(k+1) = \mathcal{M}(x(k))$, $k \geq 0$, where $\mathcal{M}$ is a min-max-plus function of dimension n.*

Subclasses of min-max-plus systems can be defined for which specific properties are known to hold. Two such subclasses, those of separated and of bipartite min-max-plus systems, will be dealt with briefly in the coming sections. The definitions are as follows.

DEFINITION 12.4 *Consider a min-max-plus system characterized by the min-max-plus function $\mathcal{M}$. If each component of $\mathcal{M}$ is either a max-plus expression or a min-plus expression only, then the system is called* separated.

If, through a possible reordering of the state components, the first n components of $\mathcal{M}$ are max-plus expressions and the last m components are min-plus expressions (by abuse of notation, the dimension of the system now is $n + m$), then, with a renaming of the state variables, we can write

$$x_i(k+1) = \max(x_1(k) + a_{i1}, \ldots, x_n(k) + a_{in}, y_1(k) + b_{i1}, \ldots, y_m(k) + b_{im}),$$
$$y_j(k+1) = \min(x_1(k) + c_{j1}, \ldots, x_n(k) + c_{jn}, y_1(k) + d_{j1}, \ldots, y_m(k) + d_{jm}),$$

for $i \in \underline{n}$ and $j \in \underline{m}$. More concisely, we can write

$$x(k+1) = (A \otimes x(k)) \oplus (B \otimes y(k)), \tag{12.6}$$
$$y(k+1) = (C \otimes' x(k)) \oplus' (D \otimes' y(k)). \tag{12.7}$$

DEFINITION 12.5 *Bipartite systems form a subclass of the class of separated systems in the sense that bipartite systems are separated systems with $A = \mathcal{E}$; that is, all elements of A are $-\infty$, and $D = \mathcal{E}'$ (i.e., all elements of D are equal to $+\infty$).*

Bipartite systems, as well as separated systems, can be symbolized by a graph with n maximizing nodes representing $x_1, \ldots, x_n$ and m minimizing nodes representing $y_1, \ldots, y_m$. Finite entries of B represent arcs from the y-nodes to the x-nodes and finite entries of C represent arcs from the x-nodes to the y-nodes. For bipartite systems there are no other arcs. The word *bipartite* indicates that there are two distinct sets of nodes with arcs from one to the other set and conversely, and no arcs between nodes of the same set. In contrast to the graph of a bipartite system, the graph of a separated system can contain arcs between x-nodes, as well as between y-nodes.

12.1.2 Eigenvalues and cycle times

Throughout this section it is assumed that $\mathcal{M}$ is an n-dimensional min-max-plus function. The following two definitions are straightforward generalizations of the notions of eigenvalue and cycle-time vector as already introduced in Part I. The notation $\mathcal{M}^p$, where p is a positive integer, refers to $\mathcal{M}$ applied p times; that is,

$$\mathcal{M}^p(x) = \underbrace{\mathcal{M}(\cdots(\mathcal{M}(x)))}_{p \text{ times}}.$$

DEFINITION 12.6 *The vector $x \in \mathbb{R}^n$ is called an eigenvector for eigenvalue $\lambda \in \mathbb{R}$ if $\mathcal{M}(x) = \lambda + x$. The vector $x \in \mathbb{R}^n$ is a periodic point of $\mathcal{M}$ with period p if it is an eigenvector of $\mathcal{M}^p$ but not of $\mathcal{M}^k$ for any $1 \leq k < p$.*

Though generalizations are possible, we restrict ourselves here to vectors and eigenvectors with finite elements only. This is in contrast to the definition of eigenvectors of max-plus matrices; see Section 2.2. Compare the notion of period in the above definition with the one of cyclicity as defined in Section 3.1.

THEOREM 12.7 *If the limit $\lim_{k\to\infty}(\mathcal{M}^k(x)/k)$ exists for some finite vector x, then it exists for all finite vectors x and the limit is independent of the initial condition x.*

Proof. Suppose $\lim_{k\to\infty}(\mathcal{M}^k(x)/k) = a$, and let $\bar{x}$ be another finite vector. Then, nonexpansiveness with respect to the supremum norm implies

$$\lim_{k\to\infty}\left\| a - \frac{\mathcal{M}^k(\bar{x})}{k} \right\|_\infty \leq \lim_{k\to\infty}\left(\left\| a - \frac{\mathcal{M}^k(x)}{k} \right\|_\infty + \left\| \frac{\mathcal{M}^k(x) - \mathcal{M}^k(\bar{x})}{k} \right\|_\infty \right)$$

$$\leq \lim_{k\to\infty}\left\| \frac{x - \bar{x}}{k} \right\|_\infty = 0.$$

$\square$

DEFINITION 12.8 *The cycle-time vector $\chi(\mathcal{M})$ of the mapping $\mathcal{M}$ is defined as $\lim_{k\to\infty}(\mathcal{M}^k(x)/k)$, whenever this limit exists.*

It will be immediately clear now that if $\mathcal{M}$ has an eigenvalue λ, then the cycle-time vector exists and equals the vector with all components equal to λ.

Write $\mathcal{M}$ once more in its conjunctive normal form, $\mathcal{M}(x) = \min_{A_i \in S}(A_i \otimes x)$, where the set S is defined as all possible $n \times n$ A matrices by taking any combination of a max-plus expression in each component of $\mathcal{M}$. (In Example 12.1.4, for instance, S consists of all four matrices given $A_i, i \in \underline{4}$.) For any $A_i \in S$ and any x, it follows that

$$\mathcal{M}(x) \leq A_i \otimes x,$$

and hence,

$$\mathcal{M}^2(x) = \mathcal{M}(\mathcal{M}(x)) \leq \mathcal{M}(A_i \otimes x) \leq A_i \otimes (A_i \otimes x) = A_i^{\otimes 2} \otimes x,$$

where the inequalities follow from the monotonicity property. Continuing, we get $\mathcal{M}^k(x) \leq A_i^{\otimes k} \otimes x$, and, for $k \to \infty$, $\chi(\mathcal{M}) \leq \chi(A_i)$. This inequality holds for any i, and thus,

$$\chi(\mathcal{M}) \leq \min_{A_i \in S} \chi(A_i). \tag{12.8}$$

This inequality provides an upper bound for $\chi(\mathcal{M})$, whenever it exists.

Remark.   Inequality (12.8) needs some extra attention since comparing vectors by means of the (scalar) ordering relation $\leq$ only provides a partial ordering. By the definition of S it follows that an $A_{i^*} \in S$ exists such that $\chi(A_{i^*}) \leq \chi(A_i)$, for all i, and this latter inequality holds componentwise. The proofs of these statements are left as an exercise (see exercise 2).

We can do the same analysis again, but now starting from the disjunctive normal form, $\mathcal{M}(x) = \max_{B_j \in T}(B_j \otimes' x)$, where the set T is defined as all possible $n \times n$ B matrices by taking any combination of a min-plus expression in each component of $\mathcal{M}$. Since $B_j \otimes' x \leq A_i \otimes x$, for any i, j combination, we obtain

$$\max_{B_j \in T} \chi(B_j) \leq \min_{A_i \in S} \chi(A_i), \tag{12.9}$$

and if $\chi(\mathcal{M})$ exists, it must have a value between these two terms. The *duality conjecture* asserts that the inequality in (12.9) can be replaced by the equality sign. For a proof of this assertion, which thus has become a truth, see, for instance, [12] and [24]. If $\mathcal{M}$ has an eigenvalue, then the proof is simple as shown by the following theorem.

THEOREM 12.9 *If $\mathcal{M}$ has an eigenvalue, then the duality conjecture holds.*

Proof. Call the eigenvalue and corresponding eigenvector λ and v, respectively, i.e., $\mathcal{M}(v) = \lambda + v$. Hence, $\chi(\mathcal{M})$ exists because $\mathcal{M}^k(v) = \lambda^{\otimes k} \otimes v = k \times \lambda + v$ for $k = 1, 2, \ldots$, implying that each component of $\chi(\mathcal{M})$ equals λ. For at least one i, $\mathcal{M}(v) = A_i \otimes v$ and $\chi(\mathcal{M}) = \chi(A_i)$. In the same way, $\chi(\mathcal{M}) = \chi(B_j)$ for some j. The two latter equalities prove the equality sign in (12.9). $\square$

12.1.3 Results on separated systems

The notation for a separated system will be the one given in (12.6) and (12.7). Please be reminded of the fact that the state is $(x^\top, y^\top)^\top$, which has size $n + m$.

THEOREM 12.10 *Assume we are given a separated system characterized by $\mathcal{M}$ and as defined by (12.6) and (12.7), with A and D being irreducible matrices and both B and C having at least one finite element. Then, the mapping $\mathcal{M}$ has an eigenvalue λ if and only if $\lambda_{\max} \leq \lambda_{\min}$, where $\lambda_{\max}$ is the eigenvalue of A (in the max-plus algebra sense) and where $\lambda_{\min}$ is the eigenvalue of D (in the min-plus algebra sense). Moreover, if $\lambda_{\max} \leq \lambda_{\min}$, then λ is unique and satisfies $\lambda_{\max} \leq \lambda \leq \lambda_{\min}$.*

At least two proofs of this theorem exist in [71] and [24]; both are rather long, and we do not give them here. However, it is easily argued that $\lambda_{\max} \leq \lambda_{\min}$ is a necessary condition for the existence of λ. If we disregard the matrices B and C, then $\mathcal{M}$ consists of two uncoupled systems,

$$x(k + 1) = A \otimes x(k), \tag{12.10}$$

$$y(k + 1) = D \otimes' y(k), \tag{12.11}$$

with the substate x growing with an average rate of $\lambda_{\max}$ (i.e., on the average $x_i(k + 1) = x_i(k) + \lambda_{\max}$ for $i \in \underline{n}$) and the substate y growing with an average rate of $\lambda_{\min}$ (on the average $y_j(k + 1) = y_j(k) + \lambda_{\min}$ for $j \in \underline{m}$). Adding the part $B \otimes y(k)$ to the right-hand side of (12.10) such as to obtain (12.6) can only further speed up the rate of x (in the sense that the time instants $x(k + 1)$ will occur later), whereas adding the term $C \otimes' x(k)$ to (12.11), so as to obtain (12.7), can only slow down the rate of y (in the sense that the time instants $y(k + 1)$ will occur sooner). Hence, if $\lambda_{\max} > \lambda_{\min}$, then the rates of the two subsystems can only grow further apart, and it will therefore be impossible for the two (sub-)states x and y to grow at an identical average rate. Hence, for the existence of a common average rate the inequality $\lambda_{\max} \leq \lambda_{\min}$ is needed.

Example 12.1.5 *A separated system is given by means of its matrices*

$$A = \begin{pmatrix} -\infty & 1 & -\infty \\ -\infty & 0 & 1 \\ 2 & 1 & 0 \end{pmatrix}, \quad B = \begin{pmatrix} 3 & 3 & -\infty \\ 3 & -\infty & -\infty \\ -\infty & -\infty & 1 \end{pmatrix},$$

$$C = \begin{pmatrix} +\infty & +\infty & 3 \\ +\infty & 3 & +\infty \\ +\infty & +\infty & 3 \end{pmatrix}, \quad D = \begin{pmatrix} +\infty & 4 & 3 \\ 6 & +\infty & +\infty \\ +\infty & 9 & 6 \end{pmatrix}.$$

The corresponding graph is shown in Figure 12.1. It is easily verified that A is irreducible and that $\lambda_{\max} = 4/3$. Similarly, D is irreducible and $\lambda_{\min} = 5$. Hence, λ must exist. One way to find λ is to use the duality conjecture, i.e., (12.9) with the equality sign. Other (numerical) approaches are mentioned in Section 12.1.5. Whatever method is used, the eigenvalue is $\lambda = 14/5$, with corresponding eigenvector $v = \frac{1}{5}(30, 28, 26, 27, 29, 27)^{\top}$.

The communication graph of the system (12.6) and (12.7) can be given. It consists of n maximizing nodes and m minimizing nodes. The finite elements of the matrices A, B, C, and D represent directed arcs. Now, the critical graph can be defined in the usual way. Toward this end one considers (12.6) and (12.7), in which an eigenvector is substituted. In each of the $n + m$ components of (12.6) and (12.7) those terms on the right-hand side that take care of the equality sign characterize a critical arc. All these critical arcs together (there are at least $n + m$) form at least one circuit, called a *critical circuit*.

For the example above, for instance, $n = 3$, $m = 3$, and the critical circuit is formed by the nodes $x_1, x_3, y_1, x_2, y_2, x_1$, visited in this order. Indeed, this circuit has average weight $14/5$. The critical graph thus constructed depends on the eigenvector chosen. The order is here first to compute an eigenvector and eigenvalue pair and then to determine the critical graph.

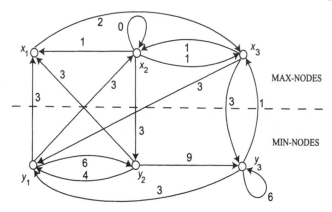

Figure 12.1: Graph corresponding to Example 12.1.5.

One could have realized beforehand that the eigenvalue should be equal to the average weight of a circuit. Since there is a finite number of circuits, one must choose out of a finite number of possibilities for λ. In general, the critical circuit is neither the slowest one (in the average sense) nor the fastest one (in the average sense). In the above example, the circuit $((y_3, x_3), (x_3, x_2), (x_2, y_2), (y_2, y_3))$ has average weight $14/4$, which is larger than $\lambda = 14/5$. Similarly, the circuit $((y_1, x_2), (x_2, x_3), (x_3, y_3), (y_3, y_1))$ has average weight $10/4$ which is smaller than λ.

Remark. Nothing has been said about the possibility of non-colinear eigenvectors. In principle, it is therefore possible that the critical circuit depends on the eigenvector chosen.

12.1.4 Results on bipartite systems

The notation for a bipartite system will be the one given in (12.6) and (12.7) with A and D nonexisting, i.e., $A = \mathcal{E}$ and $D = \mathcal{E}'$. Hence, a bipartite system is characterized by two matrices B and C, such that

$$x(k + 1) = B \otimes y(k), \qquad y(k + 1) = C \otimes' x(k). \qquad (12.12)$$

It will be assumed in this section that each row of B and each row of C contain at least one finite entry. Systems that satisfy this assumption are called *regular*, just as for max-plus algebra.

THEOREM 12.11 *Consider the regular bipartite system (12.12). If the matrix pair (B, C) is irreducible, then an eigenvalue (with corresponding eigenvector) exists.*

A matrix pair being irreducible is an extension of a single matrix being irreducible. The definition is as follows. If σ denotes a permutation of $\underline{n}$ and τ a permutation of $\underline{m}$, then the $n \times m$ matrix $W(\sigma, \tau)$ is obtained from the $n \times m$ matrix W by permuting the rows and columns of W according to σ and τ, respectively.

DEFINITION 12.12 *The matrix pair (B, C) is irreducible if no permutations σ of $\underline{n}$ and τ of $\underline{m}$ exist such that*

$$B(\sigma, \tau) = \begin{pmatrix} B_{11} & B_{12} \\ \mathcal{E} & B_{22} \end{pmatrix}, \qquad C(\tau, \sigma) = \begin{pmatrix} C_{11} & \mathcal{E}' \\ C_{21} & C_{22} \end{pmatrix},$$

where

- *the sizes of B_{ij} and C_{ji}^{T}, $i, j \in \underline{2}$, are identical (the submatrices B_{ii} and C_{jj} are not necessarily square), and*

- *B_{11} and C_{22} are regular.*

Otherwise, the pair (B, C) is called reducible.

The reader is referred to the exercises in Section 12.3 in order to show that this definition can be viewed as an extension of the definition of irreducibility for single matrices. A bipartite system characterized by a nonirreducible (i.e., reducible) matrix pair can be written as, after a possible reordering of the components of the state vector,

$$x_1(k + 1) = B_{11} \otimes y_1(k) \oplus B_{12} \otimes y_2(k), \tag{12.13}$$
$$x_2(k + 1) = \phantom{B_{11} \otimes y_1(k) \oplus} B_{22} \otimes y_2(k),$$
$$y_1(k + 1) = C_{11} \otimes' x_1(k),$$
$$y_2(k + 1) = C_{21} \otimes' x_1(k) \oplus' C_{22} \otimes' x_2(k), \tag{12.14}$$

where the vector x has been split up into two subvectors x_i, $i \in \underline{2}$, of appropriate size and similarly for y. If the original system is regular, then the individual subsystems

$$\begin{pmatrix} x_1(k + 1) \\ y_1(k + 1) \end{pmatrix} = \begin{pmatrix} B_{11} \otimes y_1(k) \\ C_{11} \otimes' x_1(k) \end{pmatrix}, \quad \begin{pmatrix} x_2(k + 1) \\ y_2(k + 1) \end{pmatrix} = \begin{pmatrix} B_{22} \otimes y_2(k) \\ C_{22} \otimes' x_2(k) \end{pmatrix}$$

are both regular bipartite systems. Suppose that both subsystems have an eigenvalue, say, λ_1 and λ_2, respectively, with $\lambda_1 > \lambda_2$. Intuitively, the average behavior of the events characterized by the time instants $x_1(k)$ and $y_1(k)$ and parameterized by k in the original model can only become slower; that is, the time instants $x_1(k + 1)$ and $y_1(k + 1)$ will occur later, due to the term $B_{12} \otimes y_2(k)$ in (12.13). Similarly, the average behavior of the events characterized by $x_2(k)$ and $y_2(k)$ in the original model can only become faster; that is, the time instants $x_2(k + 1)$ and $y_2(k+1)$ will occur sooner, due to the term $C_{21} \otimes' x_1(k)$ in (12.14). This is a plausible argument to support the statement that the average rate of growth of x_1, y_1 on the one side and of x_2, y_2 on the other will never become equal and hence, the eigenvalue for the original nonirreducible system cannot exist. Therefore, for the eigenvalue of the overall system to exist, it must be true that $\lambda_1 \leq \lambda_2$.

The existence of the eigenvalue in Theorem 12.11 depends purely on qualitative properties of the matrix pair (B, C) and not on the numerical values of the finite elements of these matrices (as long as they remain finite). Thus, one talks about the *structural* existence of an eigenvalue. By abuse of language, the expression *structural eigenvalue* is also used. A system characterized by a nonirreducible matrix pair (B, C) may or may not have an eigenvalue. The existence of the latter depends on the numerical values of the elements of the matrices B and C. In that case one speaks of a *nonstructural* eigenvalue (provided that it exists). Note that the eigenvalue of a separated system is always nonstructural.

12.1.5 Some remarks on algorithmic issues

For general min-max-plus systems, the cycle-time vector can in principle be calculated by employing the duality conjecture, i.e., $\chi(\mathcal{M}) = \min_{A_i \in S} \chi(A_i)$; see (12.8) and (12.9). The set S, however, though finite, may be very large. In [24] ideas of the policy algorithm are presented in order to speed up this approach. Recently, the policy algorithm extended to bipartite systems has been shown to work well; see [81].

For irreducible bipartite systems, the following power algorithm yields the eigenvalue and an eigenvector; see [82].

Algorithm 12.1.1

1. *Start with an arbitrary vector $x(0)$.*

2. *Iterate $x(k+1) = \mathcal{M}(x(k))$, $k = 0, 1, \ldots$, until there are integers p, q, with $p > q \geq 0$ and a finite real number c such that $x(p) = c \otimes x(q)$.*

3. *Define as eigenvalue $\lambda = c/(p - q)$ and as candidate eigenvector*

$$v = \bigoplus_{i=1}^{p-q} \lambda^{\otimes(p-q-i)} \otimes x(q + i - 1).$$

Alternatively, one can take the candidate eigenvector

$$v = \bigoplus_{i=1}^{p-q}{}' \lambda^{\otimes(p-q-i)} \otimes x(q + i - 1).$$

4. *If $\mathcal{M}(v) = \lambda \otimes v$, then v is a correct eigenvector; stop. Otherwise, start again at step 2, with $x(0) = v$ as the new initial state vector. Thus, the newly obtained quantities λ and v in step 3 do satisfy $\mathcal{M}(v) = \lambda \otimes v$.*

This algorithm even seems to work for a wider class of systems, such as, for example, separated systems as shown now by its application to Example 12.1.5. There one finds, if one starts with the zero-vector, that the subsequent states are

$$
\begin{pmatrix} 0 \\ 0 \\ 0 \\ 0 \\ 0 \\ 0 \end{pmatrix}, \begin{pmatrix} 3 \\ 3 \\ 2 \\ 3 \\ 3 \\ 3 \end{pmatrix}, \begin{pmatrix} 6 \\ 6 \\ 5 \\ 5 \\ 6 \\ 5 \end{pmatrix}, \begin{pmatrix} 9 \\ 8 \\ 8 \\ 8 \\ 9 \\ 8 \end{pmatrix}, \begin{pmatrix} 12 \\ 11 \\ 11 \\ 11 \\ 11 \\ 11 \end{pmatrix}, \begin{pmatrix} 14 \\ 14 \\ 14 \\ 14 \\ 14 \\ 14 \end{pmatrix},
$$

and a periodic behavior is obtained after five steps. So, $p = 5$, $q = 0$, and $c = 14$. Generally, there is a transient behavior (i.e., the phase from $x(0)$ to $x(q)$), but little is known about its maximum length; see, however, [50] and [81]. The algorithm now gives that $\lambda = 14/5$ is the eigenvalue and that $v = \frac{1}{5}(60, 58, 56, 57, 59, 57)^\top$ is an eigenvector. These claims for the eigenvalue and eigenvector are easily shown to be correct by substitution into $\mathcal{M}(v) = \lambda + v$. Note that the vector v and the eigenvector in Example 12.1.5 are colinear.

For separated systems another algorithm is given in [71]. Essentially, one studies $||c||_{\mathbb{P}} = \max_i c_i - \min_i c_i$, where $c_i = \mathcal{M}_i(x) - x_i$, $i \in \underline{n}$, as a function of the state x. The mapping $\mathcal{M}$ refers to the separated system under consideration. One continuously adapts the vector x in such a way that with these changes the quantity $||c||_{\mathbb{P}}$ decreases and ultimately becomes zero. The x vector for which $||c||_{\mathbb{P}} = 0$ is an eigenvector, and $c_1 = \cdots = c_n$ is the eigenvalue.

12.2 LINKS TO OTHER MATHEMATICAL AREAS

12.2.1 Link with the theory of nonnegative matrices

Consider once more the specific eigenvalue problem introduced in Chapter 0; that is,

$$\max(2 + v_1, 5 + v_2) = \lambda + v_1,$$
$$\max(3 + v_1, 3 + v_2) = \lambda + v_2.$$

By means of

$$\max(a, b) = \lim_{s \to \infty} \frac{1}{s} \ln(e^{sa} + e^{sb}), \qquad a + b = \lim_{s \to \infty} \frac{1}{s} \ln(e^{sa} e^{sb}), \quad (12.15)$$

the two scalar equations for the eigenvector can be approximated by

$$\frac{1}{s} \ln(e^{s(2+v_1)} + e^{s(5+v_2)}) = \frac{1}{s} \ln(e^{s(\lambda+v_1)}),$$
$$\frac{1}{s} \ln(e^{s(3+v_2)} + e^{s(3+v_2)}) = \frac{1}{s} \ln(e^{s(\lambda+v_2)})$$

or, equivalently, by

$$\begin{aligned} e^{2s} e^{sv_1} + e^{5s} e^{sv_2} &= e^{s\lambda} e^{sv_1}, \\ e^{3s} e^{sv_2} + e^{3s} e^{sv_2} &= e^{s\lambda} e^{sv_2}. \end{aligned} \quad (12.16)$$

The reader will have realized that e here stands for $\exp$ (of exponential). For $s \to \infty$ the approximation becomes exact in the appropriate sense. Now note that (12.16) is the eigenvalue equation for the matrix

$$A = \begin{pmatrix} e^{2s} & e^{5s} \\ e^{3s} & e^{3s} \end{pmatrix} \quad (12.17)$$

in conventional algebra, where now the eigenvalue is indicated by $e^{s\lambda}$ and the components of the eigenvector by e^{sv_i}, $i \in \underline{2}$. Since the elements of this matrix are positive, the Perron-Frobenius theorem [11] teaches us that a real and positive eigenvalue exists with a corresponding eigenvector of which the elements are real and positive also. Hence, the fact that the eigenvalue and the components of the eigenvector in (12.16), which can actually be seen as a definition of these quantities, are restricted to be positive is not a restriction.

For the sake of completeness, let us solve the eigenvalue for the matrix (12.17) in conventional algebra. Then, the eigenvalue $e^{s\lambda}$ must satisfy $\det(A - e^{s\lambda}I) = 0$, where I denotes the identity matrix in conventional linear algebra. It follows that

$$(e^{s\lambda})^2 - (e^{2s} + e^{3s})e^{s\lambda} - e^{8s} = 0,$$

which has as solutions

$$e^{s\lambda} = \frac{(e^{2s} + e^{3s}) \pm \sqrt{(e^{2s} + e^{3s})^2 + 4e^{8s}}}{2}.$$

For $s \to \infty$ one obtains for the positive eigenvalue $e^{s\lambda} = e^{4s}$ in the appropriate sense, and thus, λ of the original problem in max-plus algebra equals 4, which is in complete agreement with the results obtained in Chapter 0.

Actually, what we did above can be interpreted as the calculation of the eigenvalue for a matrix in max-plus algebra via a detour in conventional algebra. The same detour has been used to prove other properties in the theory of max-plus algebra.

12.2.2 Imbedding in nonexpansive maps

In Section 12.1.1 we encountered three properties of a min-max-plus function, namely, monotonicity, homogeneity and nonexpansiveness with respect to the l^∞-norm. These properties will briefly be indicated by the symbols M, H and N, respectively. In the literature results are given in the case where nonexpansiveness is defined by means of a different norm (specifically, the l^1-norm); however, we will confine ourselves in this section to the l^∞-norm.

THEOREM 12.13 *If the function* $\mathcal{M} : \mathbb{R}^n \to \mathbb{R}^n$ *satisfies* H, *then* M *is equivalent to* N.

In the statement of Theorem 12.13, $\mathcal{M}$ is not necessarily restricted to be a min-max-plus function. It is simply a mapping from $\mathbb{R}^n$ into $\mathbb{R}^n$ satisfying the above-mentioned properties. Such mappings (satisfying H and M, and equivalently, satisfying H and N) are called *topical*. That the min-max-plus functions form an actual subset of the set of topical functions is shown in the following example of a topical function, which is not min-max-plus; see also exercise 8.

Example 12.2.1   *Consider the mapping* $f : \mathbb{R}^n \to \mathbb{R}^n$ *symbolized by* $\ln(A \exp(\cdot))$, *where* A *is a positive matrix and both* $\ln(\cdot)$ *and* $\exp(\cdot)$ *are defined componentwise, i.e.,* $(\ln(x))_i = \ln(x_i)$ *and* $(\exp(x))_i = \exp(x_i)$. *As a specific example (with* $n = 2$ *and* A *being the matrix of (0.10)),*

$$f_1(x) = \ln(2e^{x_1} + 5e^{x_2}),$$
$$f_2(x) = \ln(3e^{x_1} + 3e^{x_2}).$$

It can be shown that any topical function can be represented as

$$\bigwedge_{i \in I} f_i, \quad \text{or as} \quad \bigvee_{j \in J} g_j,$$

where I and J are possibly uncountably infinite and where the components of f_i and g_j are max-plus and min-plus expressions, respectively. Note that if I and J are finite, then these representations are the disjunctive and conjunctive normal forms as already introduced.

One may now wonder whether the theory of eigenvalues and cycle times, as developed for min-max-plus systems in Section 12.1.2, can be carried over to topical functions. Definitions 12.6 and 12.8, as well as Theorem 12.7, are valid in the current context of nonexpansive mappings.

THEOREM 12.14 *If* $\mathcal{M} : \mathbb{R}^n \to \mathbb{R}^n$ *is nonexpansive and if* p *is the period (i.e.,* $\mathcal{M}^{p+1}(\cdot) = \mathcal{M}(\cdot)$), *then* $p \leq (2n)^n$.

For max-plus systems (or min-plus systems), it is easy to show that a tighter upper bound can be given. In contrast to max-plus systems, the cycle-time vector does not always exist for nonexpansive mappings. In [49] a counterexample to this extent has been given with $n = 3$. For $n = 1, 2$, the cycle-time vector always exists.

12.3 EXERCISES

1. Consider the two-dimensional system

$$x(k + 1) = \min(\max(x(k) + 1, y(k) + 7), \max(x(k) + 6, y(k) + 4)),$$
$$y(k + 1) = \min(\max(x(k) + 8, y(k) + 2), \max(x(k) + 3, y(k) + 5)).$$

Calculate the eigenvalue (answer: $\lambda = 5$) and show that the duality conjecture holds for this system. Rewrite the system in its disjunctive normal form, recalculate the eigenvalue, and show the correctness of the duality conjecture now starting from this representation.

2. Prove the statements made in the remark in Section 12.1.2.

3. Prove that the irreducibility of a square matrix A is equivalent to the irreducibility of the matrix pair (A, B), where B is the identity matrix in min-plus algebra.

4. Calculate, by means of the power algorithm, the eigenvalue and an eigenvector of the bipartite system characterized by

$$B = \begin{pmatrix} 2 & -3 & 6 & 2 & -11 \\ 13 & 12 & 19 & -6 & 21 \\ -10 & 8 & 14 & -5 & -16 \end{pmatrix}, \quad C = \begin{pmatrix} 18 & 8 & 4 \\ -11 & 10 & 14 \\ -8 & -4 & 4 \\ 13 & -1 & -7 \\ 4 & 7 & 0 \end{pmatrix}.$$

(Answer: $\lambda = 3$, $v = (14, 29, 14, 15, 0, 3, 4, 11)^\top$.)

5. Consider the bipartite system characterized by

$$B = \begin{pmatrix} 1 & -\infty \\ -\infty & 2 \end{pmatrix}, \quad C = \begin{pmatrix} 1 & 100 \\ 100 & 0 \end{pmatrix}.$$

Show that this system has at least two independent eigenvectors $((1, 2, 1, 1)^\top$ and $(2, 2, 2, 1)^\top$). The critical circuit as defined in Section 12.1.3 depends in principle on the eigenvector. However, show that this is not true in the current example.

6. Suppose you are asked to design a timetable of trains on a network of tracks with a real crossing, say, a crossing of an east–west line with a north–south line. On this crossing the order of events is such that the first arriving train, which could arrive from either the east or north, should pass first (and the second arriving train must possibly wait for this first train to have passed), and afterwards the trains of the east–west direction and the north–south direction must alternate. Can you come up with a min-max-plus model to allow for the inclusion of the freedom to order the trains in this way?

7. In Section 12.2.1 a detour via conventional algebra has been defined to prove some results in max-plus algebra. Is a similar detour possible to prove results in the min-max-plus algebra by considering $\min(a, b) = \lim_{s \to \infty} \frac{1}{s} \ln(e^{-sa} + e^{-sb})$ in addition to (12.15)?

8. Show that the mapping f in Example 12.2.1 is topical.

12.4 NOTES

Most of the material of Section 12.1 has been taken from [24], [71], and [85]. It can be shown that min-max-plus functions are dense in the class of topical functions in an appropriate setting; see [24]. The definitions of eigenvectors in some papers are more general than the one given here, in the sense that some elements (but not all) may be $-\infty$ or $+\infty$. This may lead to slightly different conditions on uniqueness issues, for instance. In [24], applications of min-max-plus algebra to the area of circuit theory, specifically for the clock schedule verification problem, are claimed. The notion of irreducibility of matrix pairs already shows up in [66]. No efficient algorithms are known to the authors to check the (ir)reducibility of matrix pairs of a large size. A recent paper with an algorithm to calculate the cycle time of min-max-plus systems is [22]. Survey paper [68] deals with the existence of cycle-time vectors (albeit in a different context). Paper [69] is a recent contribution toward the theory of nonexpansive maps.

Chapter Thirteen

Continuous and Synchronized Flows on Networks

So far, we have formulated timed events as discrete flows on networks. In this section, we consider a continuous version of such flows.

One possible way to define, describe, and analyze such continuous flows is by limit arguments in timed event graphs (Chapter 7). In such an approach tokens are split up into mini-tokens (say, one original token consists of N identical mini-tokens); the original corresponding place is replaced by N places in series, with one mini-token in each of them and with transitions in between. The original holding times are divided by N (firing times remain zero). A transition can fire when each of the upstream places contains at least one mini-token. In the limit, when $N \to \infty$, the result is something that is called a continuous flow, which, due to the behavior of transitions, is synchronized.

Instead, we will follow a slightly different route to introduce such flows. Many of the results of Chapter 3 also hold here, just as there are many similarities between recurrence equations and ordinary differential equations.

13.1 DATER AND COUNTER DESCRIPTIONS

Compare the dater and counter descriptions (see Section 0.5) of the same system:

$$x_i(k) = \bigoplus_{j=1}^{n} a_{ij} \otimes x_j(k - b_{ij}), \qquad i \in \underline{n}, \tag{13.1}$$

$$\kappa_i(\chi) = \bigoplus_{j=1}^{n}{}' b_{ij} \otimes \kappa_j(\chi - a_{ij}), \qquad i \in \underline{n}. \tag{13.2}$$

In these descriptions the quantities b_{ij} are natural numbers that refer to the number of (unit) delays in the counting. Quite often $b_{ij} = 1$ (for instance, after having augmented the original state vector, such that the dater equations have become a first-order recurrence equation). The quantities a_{ij}, which are real valued, refer to travel times between the nodes of the network. Quantity $x_i(k)$ refers to the time instant at which the kth event occurs; $\kappa_i(\chi)$ refers to the number of events that have occurred up to (and including) time χ.

In the parlance of Petri nets, b_{ij} is the number of tokens in the place that one passes if traveling from transition q_j to transition q_i; see Chapter 7. The quantities a_{ij} are holding times, and we assume the firing times to be zero. Note that in the Petri net interpretation of (13.1) or (13.2) there is maximally one connection, with one place, between two transitions. In the context of Petri nets, (13.1) and (13.2) can be rewritten as

$$x_i(k) = \bigoplus_{j \in \pi(i)} a_{ij} \otimes x_j(k - b_{ij}), \qquad i \in \underline{n}, \tag{13.3}$$

$$\kappa_i(\chi) = \bigoplus_{j \in \pi(i)}{}' b_{ij} \otimes \kappa_j(\chi - a_{ij}), \qquad i \in \underline{n}, \tag{13.4}$$

where, as before, the set $\pi(i)$ refers to the immediate upstream transitions of q_i. In the latter two equations, the quantities a_{ij} and b_{ij} are assumed to be finitely valued. As a reminder, the notation $\sigma(i)$, to be used later on again, refers to the set of immediate downstream transitions of q_i.

Though (13.3) and (13.4) essentially describe the same phenomena, there is a clear asymmetry between the two models. In (13.3) the delays b_{ij} are integer valued and the coefficients a_{ij} real valued, whereas in (13.4) the delays, here a_{ij}, are real valued and the coefficients b_{ij} integer valued. The extension to be made now is that it does not matter whether one prefers (13.3) or (13.4) for further analysis; both a_{ij} and b_{ij} are assumed to be nonnegative and real valued. Hence, the components of the states x and κ are real valued.

The interpretation of (13.1) and (13.2), with a_{ij} and b_{ij} real valued, is still a (strongly connected) network with n nodes (or transitions in Petri net terminology). These nodes can now fire continuously.

Example 13.1.1 *In a specific country, rosé wine is made by pouring white and red wines together. One tap delivers red wine, the other tap white wine. The separate flows of red and white come together at a transition, which mixes the incoming flows, at equal rates, instantaneously and continuously, into rosé (as long as the incoming streams do not dry up). The outgoing continuous flow of rosé is subsequently led to a bottling machine. The amount of rosé produced in this way, in liters, say, and up to a certain time χ, is $2 \times \min(\kappa_{\text{white}}(\chi), \kappa_{\text{red}}(\chi))$, where $\kappa_{\text{white}}(\chi)$ is the total amount of white wine offered up to time χ to the transition and $\kappa_{\text{red}}(\chi)$ is likewise defined. If for instance $\kappa_{\text{white}}(\chi) > \kappa_{\text{red}}(\chi)$, then part of the white wine must be stored temporarily in a buffer in order to be mixed later on when more red wine becomes available.*

For the solution of (13.3) and (13.4), initial conditions should be given. They are

$$x_j(s) \text{ for } \quad - \max_{l \in \sigma(j)} b_{lj} \leq s \leq 0,$$

$$\kappa_j(s) \text{ for } \quad - \max_{l \in \sigma(j)} a_{lj} \leq s \leq 0,$$

respectively, for $j \in \underline{n}$. For (13.3) and (13.4) to be solvable unambiguously, the conditions $a_{ij} > 0$ and $b_{ij} > 0$ are certainly sufficient but not necessary.

13.2 CONTINUOUS FLOWS WITHOUT CAPACITY CONSTRAINTS

The intensity by means of which node q_j fires at time χ is indicated by $v_j(\chi)$. Obviously, $v_j(\chi) \geq 0$. In this section there are no upper bounds on $v_j(\chi)$, i.e., the node can produce at an arbitrarily high rate. Quantity $\kappa_j(\chi)$ denotes the total amount produced by node q_j up to (and including) time χ. In order to have a handy visualization, the outgoing production is assumed to move with unit speed to the downstream nodes. An artificial length of a_{ij} for the connection between nodes q_j and q_i is subsequently assumed, such that the travel time is indeed a_{ij}. Along an arc there is a continuous flow, and its intensity will be denoted by $\phi_j(\chi, l)$, where l is the parameter indicating the exact location along an arc starting at node q_j. Node q_j may have many outgoing arcs, but the flow $\phi_j(\chi, l)$ will be the same along each of them, though it is possible that the range of l will be different for each of these arcs; see Figure 13.1. The beginning of the arc from q_j to q_i coincides with $l = 0$, and $l = a_{ij}$ coincides with its end. As long as the parameters lie in appropriate intervals, it follows that

$$\phi_j(\chi, l) = \phi_j(\chi + s, l + s), \qquad \phi_j(\chi, l) = \phi_j(\chi - l, 0) = v_j(\chi - l).$$

At time χ the total amount of material along the arc from node q_j to node q_i equals

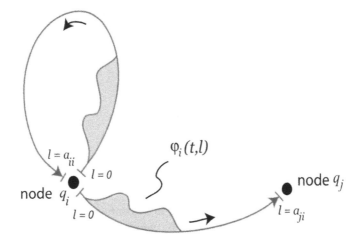

Figure 13.1: Identical continuous flows along the outgoing arcs of a node.

$$\int_0^{a_{ij}} \phi_j(\chi, s)ds. \tag{13.5}$$

The quantity b_{ij} satisfies

$$b_{ij} = \int_0^{a_{ij}} \phi_j(0, s)ds, \tag{13.6}$$

which equals the initial amount of material (a real number!) along the connection between nodes q_j and q_i. The integrands in (13.5) and (13.6) must be considered with care. It is quite possible that these integrands contain Dirac δ-functions. This will particularly happen at the end of an arc, since sometimes material must wait there to be processed by the immediate downstream transition because the other incoming arcs to the same transition have brought in less material thus far. If q_k is a downstream transition to both q_i and q_j and if $\kappa_i(\chi) < \kappa_j(\chi)$, then the flow

ϕ_j will start to build up a δ-function at $l = a_{kj}$, i.e., at the gate of q_k, at time χ. Of course, this δ-function can disappear again later on if $\kappa_i(s) > \kappa_j(s)$ for some s with $s > \chi$. Due to the fact that there are no capacity constraints, it is possible that two incoming δ-functions can be processed together and that on the outgoing arc(s) a δ-function appears which will travel with unit speed to the next transition.

The total amount of material along an arc, as expressed by (13.5), will, in general, be time (i.e., χ) dependent. However, along a circuit the total amount of material is constant.

THEOREM 13.1 *Given a circuit* $((q_{i_1}, q_{i_2}), \ldots, (q_{i_k}, q_{i_1}))$, *the total amount of material along this circuit, given by*

$$\sum_{l=1}^{k} \int_0^{a_{i_{l+1},i_l}} \phi_{i_l}(\chi, s)ds,$$

where $a_{i_{k+1},i_k} = a_{i_1,i_k}$, *is constant (i.e., does not depend on* χ).

Proof. A firing transition takes away from every incoming arc exactly as much material as it puts on each of the outgoing arcs. □

Please note that the total amount of material in the network is not necessarily constant.

DEFINITION 13.2 (Cycle mean) *Given a circuit* $\zeta = ((q_{i_1}, q_{i_2}), \ldots, (q_{i_k}, q_{i_1}))$, *its weight* $|\zeta|_w$ *and its length* $|\zeta|_1$ *are defined as*

$$|\zeta|_w = \sum_{l=1}^{k} a_{i_{l+1},i_l}, \qquad |\zeta|_1 = \sum_{l=1}^{k} b_{i_{l+1},i_l}.$$

If $|\zeta|_1 > 0$, *then the cycle mean is defined as* $|\zeta|_w / |\zeta|_1$.

This definition of the circuit length is not to be confused with the interpretation of an arc (i) having a length and (ii) along which the material flows with unit speed. The length is the total amount of material in a circuit (which coincides with the total number of tokens for the discrete flows considered in earlier chapters).

Assumption. The following will hold for the remainder of this section:

- The network is strongly connected.

- a_{ij} and b_{ij} are nonnegative along connections between transitions.

- $|\zeta|_1 > 0$ and $|\zeta|_w > 0$ for all circuits ζ.

The reason for the assumption $|\zeta|_1 > 0$, apart from the fact that it is needed in the definition of a cycle mean, is that if the total amount of material in a circuit would be zero, it will remain zero forever due to Theorem 13.1 and the transitions in this circuit will remain idle forever.

DEFINITION 13.3 *The circuits that have the maximum cycle mean are called* critical. *The corresponding cycle mean is indicated by* λ, *i.e.,* $\lambda = \max_\zeta |\zeta|_w / |\zeta|_1$.

THEOREM 13.4 *For suitably chosen initial conditions, equations (13.2) have a solution*

$$\kappa_i(\chi) = \frac{1}{\lambda} \times \chi + d_i, \tag{13.7}$$

where d_i are constants.

Proof. Suppose that solutions of the form $\kappa_i(\chi) = c_i \times \chi + d_i$ exist (in the remainder of this proof conventional multiplication ($\times$) will not be indicated explicitly anymore). They are then substituted into (13.2), leading to the identities

$$c_i\chi + d_i = \min_{j \in \pi(i)} \left\{ b_{ij} + c_j((\chi - a_{ij}) + d_j) \right\}, \qquad i \in \underline{n}. \tag{13.8}$$

For large values of χ ($\chi \to \infty$), this leads to

$$c_i = \min_{j \in \pi(i)} c_j, \qquad i \in \underline{n}.$$

Due to the assumption of the network being strongly connected, all c_i values must be equal, to be denoted here by c. Now substitute $\chi = 0$ into (13.8), resulting in

$$d_i = \min_{j \in \pi(i)} \left\{ b_{ij} - ca_{ij} + d_j \right\}, \qquad i \in \underline{n},$$

which can be written in min-plus notation as

$$d = R \otimes' d,$$

where $d = (d_1, \ldots, d_n)^\top$ and where element (i, j) of the matrix R equals $b_{ij} - ca_{ij}$, provided that $j \in \pi(i)$; otherwise, this element equals $+\infty$. The vector d is an eigenvector of R, corresponding to the eigenvalue e (in the min-plus algebra sense). The remaining question is whether c can be chosen such that R has an eigenvalue e. Due to strong connectedness again, R is irreducible, and hence, it has only one eigenvalue that equals the minimum cycle mean (recall that we are currently working in min-plus algebra; the definition of a minimum cycle mean will be obvious)

$$\min_\zeta \frac{\sum_{(j,i) \in \zeta}(b_{ij} - ca_{ij})}{|\zeta|_1} = \min_\zeta \frac{\sum_{(j,i) \in \zeta} b_{ij} - c\sum_{(j,i) \in \zeta} a_{ij}}{|\zeta|_1}. \tag{13.9}$$

If we choose $c = (\sum_{(j,i) \in \zeta^*} b_{ij})/(\sum_{(j,i) \in \zeta^*} a_{ij})$, where ζ^* is a circuit for which the minimum in (13.9) is attained, then $c = \lambda^{-1}$ and the minimum cycle mean in (13.9) becomes zero as required. Now given that (13.7) is a solution of (13.2), the initial conditions as mentioned in the statement of Theorem 13.4 are determined by this solution. This concludes the proof. $\qquad\square$

Remark. With all quantities specified as in Theorem 13.4, note that the density functions $\phi_i(t, s)$ are constant; that is, $\phi_i(t, s) = c$ for all appropriate t, s, with a possible exception of the endpoints of the corresponding arcs, where a δ-function may be present (of which the magnitude is constant with time again).

The solution given in the proof of Theorem 13.4 is, in the sense to be given, the best one. Let us concentrate on a critical circuit. During an interval of $\sum a_{ij}$ time units, where the summation is over all arcs of this circuit, any transition within this critical circuit can never produce more than an amount of $\sum b_{ij}$, the summation being again over all arcs of the circuit. In the linear solution in the statement of Theorem 13.4, any transition on the critical circuit produces exactly an amount of material equal to $\sum b_{ij}$ in each of the outgoing arcs during $\sum a_{ij}$ time units.

Example 13.2.1 *Solutions other than (13.7) may exist to (13.2). By means of this example, it will be shown that such solutions indeed exist. The solutions to be presented fluctuate in a periodic way around the linear solution obtained in Theorem 13.4. Exactly the same phenomenon has been observed for linear systems in conventional discrete max-plus (or min-plus) algebra. If one starts with an eigenvector as an initial condition, then the state behaves linearly with respect to time (one could say that this solution has period one). Other solutions are generally possible with a period larger than one and which fluctuate around the linear behavior just mentioned.*

We are given a network with $n = 2$. It is assumed that all four arcs (from transition q_i to q_j, $i, j \in \underline{2}$) exist and that

$$\lambda \overset{\text{def}}{=} \frac{a_{12} + a_{21}}{b_{12} + b_{21}} > \frac{a_{ii}}{b_{ii}}, \qquad i \in \underline{2}. \tag{13.10}$$

Try a solution of the form

$$\kappa_i(\chi) = \lambda^{-1}\chi + \alpha_i \sin(\beta(\chi - r_i)) + s_i, \qquad i \in \underline{2},$$

with $r_2 = s_2 = 0$ and $\beta > 0$, and

$$|\alpha_i|\beta < \lambda^{-1}, \qquad i \in \underline{2}. \tag{13.11}$$

The latter two conditions ensure the solutions, if they exist, to be nondecreasing. This solution is substituted into (13.2), leading to the identities

$$
\begin{aligned}
s_1 + \alpha_1 \sin(\beta(\chi - r_1)) &= \min\{-\lambda^{-1}a_{11} + \bar{b}_{11} + \alpha_1 \sin(\beta(\chi - a_{11} - r_1)), \\
&\quad -\lambda^{-1}a_{12} + b_{12} + \alpha_2 \sin \beta(\chi - a_{12})\}, \\
\alpha_2 \sin(\beta\chi) &= \min\{-\lambda^{-1}a_{21} + \bar{b}_{21} + \alpha_1 \sin(\beta(\chi - a_{21} - r_1)), \\
&\quad -\lambda^{-1}a_{22} + b_{22} + \alpha_2 \sin \beta(\chi - a_{22})\},
\end{aligned}
\tag{13.12}
$$

where $\bar{b}_{i1} = s_1 + b_{i1}$. Each minimization operation has two arguments. Assume for the moment that these arguments satisfy

$$
\begin{aligned}
-\lambda^{-1}a_{11} + s_1 + b_{11} + \alpha_1 \sin(\beta(\chi - a_{11} - r_1)) &\geq \\
-\lambda^{-1}a_{12} + b_{12} + \alpha_2 \sin \beta(\chi - a_{12}), & \\
-\lambda^{-1}a_{21} + s_1 + b_{21} + \alpha_1 \sin(\beta(\chi - a_{21} - r_1)) &\leq \\
-\lambda^{-1}a_{22} + b_{22} + \alpha_2 \sin \beta(\chi - a_{22}). &
\end{aligned}
\tag{13.13}
$$

Then, the identities (13.12) become

$$
\begin{aligned}
s_1 + \alpha_1 \sin(\beta(\chi - r_1)) &= -\lambda^{-1}a_{12} + b_{12} + \alpha_2 \sin \beta(\chi - a_{12}), \\
\alpha_2 \sin(\beta\chi) &= -\lambda^{-1}a_{21} + s_1 + b_{21} + \alpha_1 \sin(\beta(\chi - a_{21} - r_1)).
\end{aligned}
$$

These equations are indeed identities if

$$r_1 = a_{12},$$
$$\beta(a_{12} + a_{21}) = 2k\pi, \qquad k = 1, 2, \ldots, \tag{13.14}$$
$$\alpha_1 = \alpha_2,$$
$$s_1 = -\lambda^{-1}a_{12} + b_{12}, \tag{13.15}$$
$$-\lambda^{-1}a_{21} + s_1 + b_{21} = 0. \tag{13.16}$$

The quantities s_1 and λ can be uniquely solved from (13.15) and (13.16). The value of λ fortunately coincides with its value given in (13.10). The value β is determined by (13.14). Thus, it is shown that a periodic solution exists, provided that (13.11) and (13.13) are true. A simple analysis shows that this is the case for $|\alpha_1|(=|\alpha_2|)$ sufficiently small. For $k = 1$ in (13.14) we get that β equals the length of the critical circuit divided by 2π. For larger values of k we get higher harmonics. If the results of various k values are combined, then the solution becomes a Fourier series (with period $a_{12} + a_{21}$) added to the linear part.

Depending on the parameters a_{ij} and b_{ij}, one can construct a Fourier series in such a way that the ultimate solution $\kappa_i(\chi)$ becomes piecewise constant (and nondecreasing), thus leading to a real discrete flow. Points at which $\kappa_i(\chi)$ jumps refer to discrete events in the traditional sense.

CONJECTURE 13.5 *If there is a unique critical circuit, then each solution of (13.2), starting from arbitrary initial conditions, converges in finite time either to the linear solution or to a periodic solution as described in the example just given.*

A proof of this conjecture could resemble the train of thought of the proof of the discrete analogue of this theorem, which can be found in Chapter 3. A sketch of such a proof, if possible, would be as follows. Given the value $\kappa_i(\chi)$ for some χ sufficiently large, a critical path along the nodes, backward in time, is constructed according to (13.2), leading back all the way to the initial conditions. This critical path will contain a number of encirclements of the critical circuit, denoted ζ_{crit}. If now the same is done with respect to $\kappa_i(\chi + |\zeta_{\text{crit}}|_w)$, then one gets the same critical path, except for the fact that ζ_{crit} will be encircled once more than for the critical path corresponding to $\kappa_i(\chi)$. Therefore, $\kappa_i(\chi + |\zeta_{\text{crit}}|_w) = \kappa_i(\chi) + |\zeta_{\text{crit}}|_1$.

If the critical circuit is nonunique, then remarks related to the periodicity similar to those in Part I can be given for the continuous case.

13.3 CONTINUOUS FLOWS WITH CAPACITY CONSTRAINTS

The basic formulas in this section are

$$v_i(\chi) = \begin{cases} c_i & \text{if } m_{ij}(\chi) > 0 \text{ for all } j \in \pi(i), \\ c_i \oplus' \bigoplus'_{j \in \pi(i)} v_j(\chi - a_{ij}) & \text{otherwise,} \end{cases}$$

(13.17)

for $i \in \underline{n}$, and

$$\dot{m}_{ij}(\chi) = v_j(\chi - a_{ij}) - v_i(\chi),$$

(13.18)

for $i, j \in \underline{n}$. Quantity m_{ij} refers to the material sent from node j to node i that has already traveled along the whole arc (j, i) and is piled up at the "entrance" of node i, waiting to be processed. This material cannot be processed immediately, due to the constraint on the firing intensity indicated by c_i. The time derivative of m_{ij}, $\dot{m}_{ij}$, denotes the change of the size of this pile. The term $-v_i(\chi)$ refers to the processing speed, and the term $v_j(\chi - a_{ij})$ denotes the speed of new material to this pile. Quantity v_i denotes the production flow (i.e., the production per time unit) of transition q_i. Quantity $c_i > 0$ denotes the maximum firing intensity (equivalently, the capacity) of node i. The relation with the total amount produced is

$$\kappa_i(\chi) = \int_0^\chi v_i(s)\, ds.$$

In terms of the flow along the arc from q_j to q_i, the quantity $m_{ij}(\chi)$ represents the magnitude of the δ-function at the end of this arc, as described in the previous section. In the current section, traveling δ-functions along the arcs cannot arise; this is in contrast to the previous section. It is easily checked that $m_{ij}(\chi) \geq 0$ for $\chi \geq 0$, provided that the initial condition $m_{ij}(0)$ is nonnegative. It may look somewhat surprising that the quantities b_{ij} have disappeared from (13.17) and (13.18). They are, however, implicitly present in the initial conditions; that is, in order to calculate $v_i(\chi)$ for $\chi > 0$, one needs "old" values of $v_i(\chi)$ (i.e., with $\chi < 0$), and the latter functions with $\chi < 0$ are related to the quantities b_{ij}.

The equivalent expressions of (13.17) and (13.18) in the dater sense are

$$v_i(k) = \begin{cases} \dfrac{1}{c_i} & \text{if } \mu_{ij}(k) > 0 \text{ for all } j \in \pi(i), \\[2ex] \dfrac{1}{c_i} \oplus \bigoplus_{j \in \pi(i)} v_j(k - b_{ij}) & \text{otherwise,} \end{cases} \tag{13.19}$$

for $i \in \underline{n}$, and

$$\dot{\mu}_{ij}(k) = v_j(k - b_{ij}) - v_i(k), \tag{13.20}$$

for $i, j \in \underline{n}$. The reader is invited to give a meaning to the quantities μ_{ij}.

THEOREM 13.6 *Along a circuit the total amount of material is constant.*

The proof is identical to that of Theorem 13.1.

In the following theorem, the quantity λ appears again. As in Section 13.2, it is equal to the maximum cycle mean. The definition of the cycle mean, however, must now be slightly adapted.

DEFINITION 13.7 *Given a circuit $\zeta = ((q_{i_1}, q_{i_2}), \ldots, (q_{i_k}, q_{i_1}))$, the weight $|\zeta|_w$ and the length $|\zeta|_l$ are defined as*

$$|\zeta|_w = \sum_{l=1}^k a_{i_{l+1}, i_l}, \qquad |\zeta|_l = \sum_{l=1}^k b_{i_{l+1}, i_l} + m_{i_{l+1}, i_l}(0).$$

If $|\zeta|_l > 0$, then the cycle mean of ζ is defined as $|\zeta|_w / |\zeta|_l$.

It is understood here that the function $\phi(0, s)$ is a real one, that is, it no longer contains any δ-function and the b_{ij}'s are defined as in (13.6). Here also the functions $\phi_j(0, s)$ and the quantities b_{ij} are related as given in (13.6), except for possible concentrations of material at the end of the arc as just explained. Because of the capacity constraints, $\phi(\chi, s)$ will also be a real function for $\chi > 0$.

THEOREM 13.8 *For appropriately chosen initial conditions, equations (13.17) have a solution $\kappa_i(\chi) = \frac{1}{\lambda}\chi + \bar{d}_i$, where $\overline{\lambda} = \max(c_1^{-1}, \ldots, c_n^{-1}, \lambda)$ and where $\bar{d}_i$ are constants.*

Proof. If $\lambda > \max_i c_i^{-1}$ (equivalently, $\lambda^{-1} < \min_i c_i$), then the proof is identical to the one of Theorem 13.4 because none of the constraints is active. If $\lambda^{-1} \geq \min_i c_i$, then the assertion of the theorem follows from direct substitution of the proposed solution into (13.17).   □

13.4 EXERCISES

1. Show that equations (13.17) are identical to (13.2) as c_i tends to ∞.

2. Consider a network with three nodes. The transportation times are $a_{21} = 1$, $a_{12} = 1$, $a_{13} = 1$, $a_{31} = 4$, $a_{32} = 3$, and $a_{33} = 2$. The time durations that have not been mentioned refer to nonexisting arcs. The capacity constraints are $c_1 = 2$, $c_2 = 3$, and $c_3 = 4$. The initial values of the flows, $\phi_i(0, s)$, $0 \leq s \leq a_{ji}$, for $i = 1, 2, 3$ and the appropriate downstream nodes q_j, are piecewise constant;

$$\phi_1(s) = 1, \quad \text{for } 0 \leq s \leq 1; \qquad \phi_1(s) = 2, \quad \text{for } 1 < s \leq 2;$$
$$\phi_1(s) = 3, \quad \text{for } 2 < s \leq 3; \qquad \phi_1(s) = 1, \quad \text{for } 3 < s \leq 4;$$
$$\phi_2(s) = 1, \quad \text{for } 0 \leq s \leq 1; \qquad \phi_2(s) = 2, \quad \text{for } 1 < s \leq 2;$$
$$\phi_2(s) = 2, \quad \text{for } 2 < s \leq 3; \qquad \phi_3(s) = 1, \quad \text{for } 0 \leq s \leq 1;$$
$$\phi_3(s) = 1, \quad \text{for } 1 < s \leq 2.$$

These functions only have to be considered up to the end of the appropriate arc. At the end of the arcs there are δ-functions with magnitudes

$$m_{21}(0) = 4, \qquad m_{12}(0) = 1, \qquad m_{31}(0) = 1,$$
$$m_{13}(0) = 2, \qquad m_{32}(0) = 0, \qquad m_{33}(0) = 3.$$

The functions ϕ together with these δ-functions can be viewed as a picture taken of the network at time zero. Evaluate the solution $\kappa(\chi)$ of this network for $\chi \geq 0$. Show the following:

- One of the capacity constraints determines the overall speed.

- Only the linear solution exists (in other words, there are no periodic solutions).

- The transient behavior lasts 7 time units.

3. Rephrase and prove Theorem 13.4 in terms of $x_i(k)$, i.e., in the max-plus algebra setting.

4. Suppose that it would be possible to have real numbers of trains, such as $1\frac{2}{3}$ or $\sqrt{7}$ trains. Since under this new rule trains can be arbitrarily small, we may remodel train models as continuous flow models. Given the track layout of Figure 0.1, with the total number of trains being four, how would you split up these trains as a continuous flow on the network so as to get the smallest possible eigenvalue? Your answer should be a constant train density of $4/13$, i.e., $4/13$ train per unit of travel time along each track, leading to $\lambda = 13/4$. Note that in more general networks the λ thus obtained is a lower bound for the similar quantity to be obtained when dealing with integer numbers of trains.

5. Give a meaning to (13.19) and (13.20). Why have the quantities a_{ij} (seemingly) disappeared from this formulation?

13.5 NOTES

This chapter closely follows [72]. Other approaches, with different modeling features to continuous Petri nets, exist. The reader is referred to [2] and [32].

One of the authors often used Example 13.1.1 to jokingly explain about continuous flows, especially in France. Great was his surprise when a colleague from Germany told him that one particular German rosé wine, with the name Schiller, is indeed produced by pouring red and white wines together. In addition, he was told that the name did not derive from the famous poet F. von Schiller, but from the *schillernden Farben* (sparkling color) of the wine. Recently, a French colleague admitted that the French Champagne Rosé can also be produced by adding red wine to classic Champagne, according to strict rules, of course.

Bibliography

[1] Akian, M., J.-P. Quadrat, and M. Viot. "Duality between probability and optimization." In J. Gunawardena, ed., *Idempotency*, Publications of the Newton Institute, Cambridge University Press, Cambridge, U.K. 1998.

[2] Alla, H., and R. David. "Continuous and hybrid Petri nets." *Journal of Circuits, Systems and Computers*, 8(1):159–188, 1998.

[3] Ayhan, H., and F. Baccelli. "Expansions for joint Laplace transforms for stationary waiting times in (max,+)-linear systems with Poisson input." *Queuing Systems: Theory and Applications*, 37:291–328, 2001.

[4] Baccelli, F., and M. Canales. "Parallel simulation of stochastic Petri nets using recurrence equations." *ACM Transactions on Modeling and Computer Simulation*, 3:20–41, 1993.

[5] Baccelli, F., G. Cohen, G. J. Olsder, and J.-P. Quadrat. *Synchronization and Linearity*. John Wiley and Sons, New York, 1992. Text can now be downloaded from the Web site http://www-rocq.inria.fr/metalau/cohen/SED/book-online.html.

[6] Baccelli, F., and D. Hong. "Analytic expansions of (max,+) Lyapunov exponents." *Annals of Applied Probability*, 10:779–827, 2000.

[7] Baccelli, F., and P. Konstantopoulos. "Estimates of cycle times in stochastic Petri nets." In I. Karatzas, ed., *Proceedings of the Workshop on Stochastic Analysis*, Lecture Notes in Control and Information Science 177, Springer-Verlag, Berlin, 1992; 1–20.

[8] Baccelli, F., and Z. Liu. "Comparison properties of stochastic decision free Petri nets." *IEEE Transactions on Automatic Control*, 37:1905–1920, 1992.

[9] Baccelli, F., and Z. Liu. "On a class of stochastic evolution equations." *Annals of Probability*, 20:350–374, 1992.

[10] Baccelli, F., and V. Schmidt. "Taylor series expansions for Poisson-driven (max,+)-linear systems." *Annals of Applied Probability*, 6:138–185, 1996.

[11] Bapat, R. B., and T. E. S. Raghavan. *Nonnegative Matrices and Applications*. Cambridge University Press, Cambridge, U.K., 1997.

[12] Bewley, T., and E. Kohlberg. "The asymptotic solution of a recursive equation occurring in stochastic games." *Mathematics of Operations Research*, 1(4):321–336, 1976.

[13] Boom, T. J. J. van den, and B. De Schutter. "Model predictive control for perturbed max-plus-linear systems: A stochastic approach." *International Journal of Control*, 77:302–309, 2004.

[14] Bougerol, P., and J. Lacroix. *Products of Random Matrices with Applications to Schrödinger Operators*. Birkhäuser, Boston, 1985.

[15] Braker, J. G. "Max-algebra modelling and analysis of time-dependent transportation networks." In *Proceedings of the First European Control Conference*, Hermes, Grenoble, France, 1991, 1831–1836.

[16] Braker, J. G. *Algorithms and Applications in Timed Discrete Event Systems.* PhD thesis, Delft University of Technology, The Netherlands, 1993.

[17] Braker, J. G. and G. J. Olsder. "The power algorithm in the max algebra." *Linear Algebra and Its Applications*, 182:67–89, 1993.

[18] Brauer, A. "On a problem of partitions." *American Journal of Mathematics*, 64:299–312, 1942.

[19] Brilman, M., and J. Vincent. "Dynamics of synchronized parallel systems." *Communications in Statistics, Stochastic Models*, 13:605–617, 1997.

[20] Cassandras, C. G. and S. Lafortune. *Introduction to Discrete Event Systems.* Kluwer Academic Publishers, Boston, 1999.

[21] Cechlárová, K. "A note on unsolvable systems of max-min (fuzzy) equations." *Linear Algebra and Its Applications*, 310:123–128, 2000.

[22] Cheng, Y., and D.-Z. Zheng. "A cycle time computing algorithm and its application in the structural analysis of min-max systems." *Discrete Event Dynamic Systems*, 14:5–30, 2004.

[23] Cochet-Terrasson, J., G. Cohen, S. Gaubert, M. McGettrick, and J.-P. Quadrat. "Numerical computation of spectral elements in max-plus-algebra." In *Proceedings of the IFAC conference on Systems Structure and Control*, IRCT, Nantes, France, 1998, 699–706.

[24] Cochet-Terrasson, J., S. Gaubert, and J. Gunawardena. "A constructive fixed point theorem for min-max functions." *Dynamics and Stability of Systems*, 14(4):407–433, 1999.

[25] Cohen, G. "Residuation and applications." In *Algèbres Max-Plus et Application en Informatique et Automatique, École de Printemps d'Informatique Théorique*, INRIA, Rocquencourt, France, 1998, 203–233.

[26] Cohen, G. "Two-dimensional domain representation of timed event graphs." In *Algèbres Max-Plus et Application en Informatique et Automatique, École de Printemps d'Informatique Théorique*, INRIA, Rocquencourt, France, 1998, 235–258.

[27] Cohen, G., D. Dubois, J.-P. Quadrat, and M. Viot. *Analyse du Comportement Périodique de Systèmes de Production Par la Théorie des Dioides.* Rapport de Recherche 191, INRIA, Le Chesnay, France, 1983.

[28] Cohen, G., D. Dubois, J.-P. Quadrat, and M. Viot. "A linear system-theoretic view on discrete-event processes and its use for performance evaluation." *IEEE Transactions on Automatic Control*, 35:210–220, 1985.

[29] Cottenceau, B., L. Hardouin, J.-L. Boimond, and J.-L. Ferrier. "Model reference control for timed event graphs in dioids." *Automatica*, 37:1451–1458, 2001.

[30] Cuninghame-Green, R. A. "Describing industrial processes with interference and approximating their steady-state behaviour." *Operational Research Quarterly*, 13:95–100, 1962.

[31] Cuninghame-Green, R. A. *Minimax Algebra*, Lecture Notes in Economics and Mathematical Systems, Vol. 166. Springer-Verlag, Berlin, 1979.

[32] David, R., and H. Alla. *Discrete, Continuous, and Hybrid Petri Nets.* Springer-Verlag, 2004.

[33] Denardo, E. V., and B. L. Fox. "Multichain Markov renewal programs." *SIAM Journal on Applied Mathematics*, 16(3):468–487, 1968.

[34] De Schutter, B. "On the ultimate behavior of the sequence of consecutive powers of a matrix in the max-plus algebra." *Linear Algebra and Its Applications*, 307:103–117, 2000.

[35] Egmond, R.J. van. *Propagation of Delays in Public Transport*. Technical Report No. 98-39, Faculty of Technical Mathematics and Informatics, Delft University of Technology, The Netherlands, 1998.

[36] Gaubert, S. "Methods and applications of (max,+)-linear algebra." In *Proceedings of the STACS' 1997*, Lecture Notes in Computer Science, Vol. 1200. Springer-Verlag, Berlin (this report can be accessed via the Web at `http://www.inria.fr/rrrt/rr-3088.html`).

[37] Gaubert, S. *Introduction aux Systèmes Dynamiques à Événements Discrets*. Lecture Notes DEA ENSMP, Paris, 1999.

[38] Gaubert, S., P. Butkovič, and R. Cuninghame-Green. "Minimal (max,+) realization of convex sequences." *SIAM Journal on Control and Optimization*, 36:137–147, 1998.

[39] Gaubert, S., and D. Hong. *Series Expansions of Lyapunov Exponents and Forgetful Monoids*. Technical Report No. 3971, INRIA, Rocquencourt, France, 2000.

[40] Gaubert, S., and J. Mairesse. "Task resource systems and (max,+) automata." In J. Gunawardena, ed., *Idempotency*, Publications of the Newton Institute, Cambridge University Press, Cambridge, U.K. 1998.

[41] Gaubert, S., and J. Mairesse. "Asymptotic analysis of heaps of pieces and application to timed Petri nets." In *PNPM'99*, IEEE-CS Press, Saragoza, Spain, 1999.

[42] Gaubert, S., and J. Mairesse. "Modeling and analysis of timed Petri nets using heaps of pieces." *IEEE Transactions on Automatic Control*, 44:683–698, 1999.

[43] Giffler, B. "Scheduling general production systems using schedule algebra." *Naval Research Logistics Quarterly*, 10:237–255, 1963.

[44] Golan, J. S. *Semirings and Their Applications*. Kluwer Academic Publishers, Dordrecht, 1999.

[45] Gondran, M. and M. Minoux. *Graphes Dioides et Semi-Anneaux, Nouveaux Modèles et Algorithmes*. Eyrolles, Paris, 2002.

[46] Goverde, R. M. P., and M. A. Odijk. "Performance evaluation of network timetables using PETER." In J. Allan, E. Andersson, C. A. Brebbia, R. J. Hill, G. Sciutto, and S. Sone, eds., *Computers in Railways VIII*. WIT Press, Southampton, Mass., 2002.

[47] Goverde, R. M. P., and G. Soto y Koelemeijer. *Performance Evaluation of Periodic Railway Timetables: Theory and Practice*. Trail Studies in Transportation Sciences No. s2000/2, Delft University of Technology, 2000.

[48] Gunawardena, J., ed. *Idempotency*. Publications of the Newton Institute, Cambrigde University Press, Cambridge, U.K., 1998.

[49] Gunawardena, J., and M. S. Keane. *On the Existence of Cycle Times for Some Nonexpansive Maps*. Technical Report, BRIMS, Hewlett-Packard Labs, Bristol, U.K., 1995.

[50] Hartmann, M., and C. Arguelles. "Transience bounds for long walks." *Mathematics of Operations Research*, 24:414–439, 1999.

[51] Heidergott, B. "A characterization for (max,+)-linear queueing systems." *Queuing Systems: Theory and Applications*, 35:237–262, 2000.

[52] Heidergott, B., and R. E. de Vries. "Towards a control theory for transportation networks." *Discrete Event Dynamic Systems*, 11:371–398, 2001.

[53] Hennion, B. "Limit theorems for products of positive random matrices." *Annals of Applied Probability*, 25:1545–1587, 1997.

[54] Howard, R. A. *Dynamic Programming and Markov Processes*. MIT Press, Cambridge, Mass., 1960.

[55] Karp, R. "A characterization of the minimum cycle mean in a digraph." *Discrete Mathematics*, 23:309–311, 1978.

[56] Kingman, J. F. C. "Subadditive ergodic theory." *Annals of Probability*, 1:883–909, 1973.

[57] Kolokoltsov, V. N., and V.P. Maslov. *Idempotent Analysis and Its Applications*. Kluwer Academic Publishers, Dordrecht, 1997.

[58] Kort, A. de, B. Heidergott, and H. Ayhan. "A probabilistic (max,+) approach for determining railway infrastructure capacity." *European Journal of Operational Research*, 148:644–661, 2002.

[59] Le Boudec, J.-Y., and P. Thiran. *Network Calculus: A Theory of Deterministic Queuing Systems for the Internet*, Lecture Notes in Computer Science, Vol. 2050, Springer-Verlag, Berlin, 2001.

[60] Loynes, R. "The stability of queues with non-independent inter-arrival and service times." *Mathematical Proceedings of the Cambridge Philosophical Society*, 58:497–520, 1962.

[61] Mairesse, J. *A Graphical Representation of Matrices in the (max,+) Algebra*. Technical Report PR-2078, INRIA, Sophia Antipolis, France, 1993.

[62] Mairesse, J. "A graphical approach to the spectral theory in the (max,+) algebra." *IEEE Transactions on Automatic Control*, 40:1783–1789, 1995.

[63] Mairesse, J. "Products of irreducible random matrices in the (max,+) algebra." *Advances in Applied Probability*, 29:444–477, 1997.

[64] Maslov, V. P., and S. N. Samborskii. *Idempotent Analysis*. American Mathematical Society, Providence, R.I., 1992.

[65] McEneaney, W. M. "Max-plus eigenvector methods for nonlinear H_∞ problems: error analysis." *SIAM Journal on Control and Optimization*, 43:379–412, 2004.

[66] Menon, M. V. "Some spectral properties of an operator associated with a pair of nonnegative matrices." *Transactions of the American Mathematical Society*, 132:369–376, 1968.

[67] Murata, T. "Petri nets: Properties, analysis and applications." *Proceedings of the IEEE*, 77:541–580, 1989.

[68] Neyman, A. "Stochastic games and nonexpansive maps." In *Stochastic Games and Applications*, NATO Science, Series C, Mathematical and Physical Sciences, Vol. 570. Kluwer Academic Publishers, Dordrecht, 2003, 397–415.

[69] Nussbaum, R. D. and S. M. Verduyn Lunel. "Asymptotic estimates for the periods of periodic points of non-expansive maps." *Ergodic Theory and Dynamical Systems*, 23:1199–1226, 2003.

[70] Olsder, G. J. "Performance analysis of data driven networks." In J. McCanny et al., eds., *Systolic Array Processors*, Prentice Hall, Upper Saddle River, N.J., 1989, 33–41.

[71] Olsder, G. J. "Eigenvalues of dynamic min-max systems." *Discrete Event Dynamic Systems*, 1:177–207, 1991.

[72] Olsder, G. J. "Synchronized continuous flow systems." In S. Balemi, P. Kozak, and R. Smedinga, eds., *Discrete Event Systems: Modelling and Control*, Birkhäuser, Basel, 1993, 113–124.

[73] Olsder, G. J. and A. F. de Kort. "Discrete events: Time tables, capacity questions and planning issues for railway systems." In W. Gong and L. Shi, eds., *Modelling, Control, and Optimization of Complex Systems*, Kluwer Academic Publishers, Dordrecht, 2002, 237–240.

[74] Olsder, G. J., J. A. C. Resing, R. E. de Vries, M. S. Keane, and G. Hooghiemstra. "Discrete event systems with stochastic processing times." *IEEE Transactions on Automatic Control*, 35:299–302, 1990.

[75] Olsder, G. J., C. Roos, and R. J. van Egmond. "An efficient algorithm for critical circuits and finite eigenvectors in the max-plus algebra." *Linear Algebra and Its Applications*, 295:231–240, 1999.

[76] Olsder, G. J., and R. E. de Vries. "On an analogy of minimal realizations in conventional and discrete event dynamic systems." In P. Varaiya and A. B. Kurzhanskii, eds., *Proceedings of the IIASA Workshop on Discrete Event Systems, Sopron, 1987* Lecture Notes in Control and Information Sciences No. 103, Springer-Verlag, Berlin, 157–162, 1989.

[77] Ortec Consultants BV. *Manual PETER*, Gouda, The Netherlands 2001.

[78] Petri, C. A. *Kommunikation mit Automaten*. PhD thesis, Institut für Instrumentelle Mathematik, Bonn, 1962.

[79] Resing, J. A. C., R. E. de Vries, G. Hooghiemstra, M. S. Keane, and G. J. Olsder. "Asymptotic behavior of random discrete event systems." *Stochastic Processes and their Applications*, 36:195–216, 1990.

[80] Schrijver, A., and A. Steenbeek. "Dienstregelingontwikkeling voor Railned (Timetable construction for Railned)" (in Dutch). Technical Report, CWI, Amsterdam, 1994.

[81] Soto y Koelemeijer, G. *On the Behaviour of Classes of Min-Max-Plus Systems*. PhD thesis, Delft University of Technology, The Netherlands, 2003.

[82] Subiono. *On Classes of Min-Max-Plus Systems and Their Applications*. PhD thesis, Delft University of Technology, The Netherlands, 2000.

[83] Subiono, and J. W. van der Woude. "Power algorithms for (max,+)- and bipartite (min,max,+)-systems." *Discrete Event Dynamic Systems*, 10:369–389, 2000.

[84] Vries, R. E. de. *On the Asymptotic Behavior of Discrete Event Systems*. PhD thesis, Delft University of Technology, The Netherlands, 1992.

[85] Woude, J. W. van der, and Subiono. "Condition for the structural existence of an eigenvalue of a bipartite (min,max,+) system." *Theoretical Computer Science*, 293:13–24, 2003.

List of Symbols

The bold number(s) at the end of each line refer(s) to the page where the symbol is introduced or used in an alternative way.

$\mathbb{N} = \{0, 1, 2, \ldots\}$ the set of natural numbers, **14**

$\mathbb{R}$ the set of real numbers, **13**

$[A]_{ij}$ element (i, j) of matrix A; also denoted by a_{ij}, **17**

$A^\top$ the transpose of matrix A; that is, $[A^\top]_{ij} = a_{ji}$, **19**

$\oplus$ the operation max, or maximization, **13**

$\oplus'$ the operation min, or minimization, **16**

$\otimes$ plus, or addition, **13**

ε the zero element in max-plus algebra; numerical value is $\varepsilon = -\infty$, **13**, **39**

ε' the zero element in min-plus algebra; numerical value is $\varepsilon' = +\infty$, **16**

$\mathcal{E}(n, m)$ the $n \times m$ matrix with all elements equal to ε, **18**

$\mathcal{E}'(n, m)$ the $n \times m$ matrix with all elements equal to ε', **180**

$E(n, m)$ the $n \times m$ matrix with element e on the diagonal and ε elsewhere, **18**

$\mathbf{u}$ the unit vector; numerical value is $\mathbf{u} = (0, \ldots, 0)^\top$, **19**

$\mathbf{u}[\mu] = \mu \otimes \mathbf{u}$ the vector with elements equal to $\mu \in \mathbb{R}_{\max}$, **59**

e the unit in max-plus and min-plus algebra; numerical value is $e = 0$, **13**, **16**

e_j the jth base vector of $\mathbb{R}_\varepsilon^n$ with jth element zero and all other elements equal to ε, **19**

$\lceil a \rceil$ the smallest integer greater than or equal to $a \in \mathbb{R}$, **131**

A_τ the matrix A with τ subtracted from every element: $[A_\tau]_{ij} = a_{ij} - \tau$, **39**, **61**

A^* the formal power series $A^* = \bigoplus_{k \geq 0} A^{\otimes k}$, **42**

A^+ the formal power series $A^+ = \bigoplus_{k \geq 1} A^{\otimes k}$, **31**

$[B]_{.k}$ the kth column of matrix B, **39**, **74**

$V(A, \mu)$ the eigenspace of matrix A for the eigenvalue μ, **36**

$V(A)$ the eigenspace of matrix A in the case where the eigenvalue is known and unique, **36**

$\lambda = \lambda(A)$ the eigenvalue of matrix A, and in the stochastic setup, the Lyapunov exponent of $\{A(k) : k \in \mathbb{N}\}$, **36**, **173**

λ^{top} the top Lyapunov exponent of $\{A(k) : k \in \mathbb{N}\}$, **170**

λ^{bot} the bottom Lyapunov exponent of $\{A(k) : k \in \mathbb{N}\}$, **170**

$\sigma = \sigma(A)$ the cyclicity of matrix A, **50**

$\sigma_{\mathcal{G}} = \sigma_{\mathcal{G}(A)}$ the cyclicity of $\mathcal{G}(A)$, **33**

$t(A)$ the transient time of matrix A, **55**

$\mathbb{R}_{\max}$ the set $\mathbb{R} \cup \{-\infty\}$, **13**

$\mathbb{R}_{\min}$ the set $\mathbb{R} \cup \{+\infty\}$, **16**

$\mathcal{R}_{\max}$ the structure $(\mathbb{R}_{\max}, \oplus, \otimes, \varepsilon, e)$, **13**

$\mathcal{R}_{\min}$ the structure $(\mathbb{R}_{\min}, \oplus', \otimes, \varepsilon', e)$, **16**

$\underline{n}$ the set $\{1, \ldots, n\}$ for $n \in \mathbb{N} \setminus \{0\}$, **17**

$\mathcal{G}(A)$ the communication graph of matrix A, **28**

$\mathcal{N}(A)$ the set of nodes of $\mathcal{G}(A)$, **28**

$\mathcal{D}(A)$ the set of arcs of $\mathcal{G}(A)$, **28**

$\mathcal{G}^c(A)$ the critical graph of matrix A, **38**

$\mathcal{N}^c(A)$ the set of nodes of $\mathcal{G}^c(A)$, **38**

$\mathcal{D}^c(A)$ the set of arcs of $\mathcal{G}^c(A)$, **38**

$\pi(i)$ the set of direct predecessors of node i, **33**

$\pi^+(i)$ the set of all predecessors of node i, **33**

$\pi^*(i)$ the set $\pi^+(i) \cup \{i\}$, **33**

$\sigma(i)$ the set of direct successors of node i, **33**

$\sigma^+(i)$ the set of all successors of node i, **33**

$\sigma^*(i)$ the set $\sigma^+(i) \cup \{i\}$, **34**

$C(A)$ the set of all elementary circuits in $\mathcal{G}(A)$, **38**

$|p|_l$ the length of path p, **28**

$|p|_w$ the weight of path p, **29**

$i\mathcal{R}j$ node j is reachable from node i, **31**

$i\mathcal{C}j$ node j communicates with node i: $i\mathcal{R}j$ as well as $j\mathcal{R}i$, with $i\mathcal{C}i$ always true, **31**

$\overline{x}$ the equivalence class of vectors that are colinear to x, **24**

$\| \cdot \|_{\mathbb{P}}$ the projective norm, **80**

(η, v) a generalized eigenmode, **58**

γ^{top} the maximal entry of vector γ, **60**

$\|v\|_\infty$ the supremum norm of vector v, **56**

$\|A\|_{\max}$ the maximal finite element of matrix A, **168**

$\|A\|_{\min}$ the minimal finite element of matrix A, **168**

Index